Everyday
Mathematics®

The University of Chicago School Mathematics Project

STUDENT MATH JOURNAL

VOLUME 2

Mc
Graw
Hill
Education

The University of Chicago School Mathematics Project

Max Bell, Director, *Everyday Mathematics* First Edition; James McBride, Director, *Everyday Mathematics* Second Edition; Andy Isaacs, Director, *Everyday Mathematics* Third, CCSS, and Fourth Editions; Amy Dillard, Associate Director, *Everyday Mathematics* Third Edition; Rachel Malpass McCall, Associate Director, *Everyday Mathematics* CCSS and Fourth Editions; Mary Ellen Dairyko, Associate Director, *Everyday Mathematics* Fourth Edition

Authors
Jean Bell, Max Bell, John Bretzlauf, Amy Dillard, Robert Hartfield, Andy Isaacs, James McBride, Rachel Malpass McCall, Kathleen Pitvorec, Peter Saecker

Fourth Edition Grade 1
Team Leader
Rachel Malpass McCall

Writers
Meg Schleppenbach Bates, Kate Berlin, Sarah R. Burns, Gina Garza-Kling, Linda M. Sims

Open Response Team
Catherine R. Kelso, Leader; Kathryn M. Rich

Differentiation Team
Ava Belisle-Chatterjee, Leader; Martin Gartzman, Anne Sommers

Digital Development Team
Carla Agard-Strickland, Leader; John Benson, Gregory Berns-Leone, Juan Camilo Acevedo

Virtual Learning Community
Meg Schleppenbach Bates, Cheryl G. Moran, Margaret Sharkey

Technical Art
Diana Barrie, Senior Artist; Cherry Inthalangsy

UCSMP Editorial
Lila K.S. Goldstein, Senior Editor; Rachel Jacobs, Kristen Pasmore, Delna Weil

Field Test Coordination
Denise A. Porter

Field Test Teachers
Mary Alice Acton, Katrina Brown, Pamela A. Chambers, Erica Emmendorfer, Lara Galicia, Heather A. Hall, Jeewon Kim, Nicole M. Kirby, Vicky Kudwa, Stephanie Merkle, Sarah Orlowski, Jenny Pfeiffer, LeAnita Randolph, Jan Rodgers, Mindy Smith, Kellie Washington

Digital Field Test Teachers
Colleen Girard, Michelle Kutanovski, Gina Cipriani, Retonyar Ringold, Catherine Rollings, Julia Schacht, Christine Molina-Rebecca, Monica Diaz de Leon, Tiffany Barnes, Andrea Bonanno-Lersch, Debra Fields, Kellie Johnson, Elyse D'Andrea, Katie Fielden, Jamie Henry, Jill Parisi, Lauren Wolkhamer, Kenecia Moore, Julie Spaite, Sue White, Damaris Miles, Kelly Fitzgerald

Contributors
William B. Baker, John Benson, Jeanine O'Nan Brownell, Andrea Cocke, Jeanne Mills DiDomenico, Rossita Fernando, James Flanders, Lila K.S. Goldstein, Allison M. Greer, Brooke A. North, Penny Williams

Center for Elementary Mathematics and Science Education Administration
Martin Gartzman, Executive Director; Meri B. Forhan, Jose J. Fragoso, Jr., Regina Littleton, Laurie K. Thrasher

External Reviewers
The *Everyday Mathematics* authors gratefully acknowledge the work of the many scholars and teachers who reviewed plans for this edition. All decisions regarding the content and pedagogy of *Everyday Mathematics* were made by the authors and do not necessarily reflect the views of those listed below.

Elizabeth Babcock, California Academy of Sciences; Arthur J. Baroody, University of Illinois at Urbana-Champaign and University of Denver; Dawn Berk, University of Delaware; Diane J. Briars, Pittsburgh, Pennsylvania; Kathryn B. Chval, University of Missouri–Columbia; Kathleen Cramer, University of Minnesota; Ethan Danahy, Tufts University; Tom de Boor, Grunwald Associates; Louis V. DiBello, University of Illinois at Chicago; Corey Drake, Michigan State University; David Foster, Silicon Valley Mathematics Initiative; Funda Gönülateş, Michigan State University; M. Kathleen Heid, Pennsylvania State University; Natalie Jakucyn, Glenbrook South High School, Glenview, IL; Richard G. Kron, University of Chicago; Richard Lehrer, Vanderbilt University; Susan C. Levine, University of Chicago; Lorraine M. Males, University of Nebraska-Lincoln; Dr. George Mehler, Temple University and Central Bucks School District, Pennsylvania; Kenny Huy Nguyen, North Carolina State University; Mark Oreglia, University of Chicago; Sandra Overcash, Virginia Beach City Public Schools, Virginia; Raedy M. Ping, University of Chicago; Kevin L. Polk, Aveniros LLC; Sarah R. Powell, University of Texas at Austin; Janine T. Remillard, University of Pennsylvania; John P. Smith III, Michigan State University; Mary Kay Stein, University of Pittsburgh; Dale Truding, Arlington Heights District 25, Arlington Heights, Illinois; Judith S. Zawojewski, Illinois Institute of Technology

Note
Too many people have contributed to earlier editions of *Everyday Mathematics* to be listed here. Title and copyright pages for earlier editions can be found at http://everydaymath.uchicago.edu/about/ucsmp-cemse/.

www.everydaymath.com

Send all inquiries to:
McGraw-Hill Education
STEM Learning Solutions Center
8787 Orion Place
Columbus, OH 43240

ISBN: 978-0-02-143081-9
MHID: 0-02-143081-0

Printed in the United States of America.

16 LMN 21

Contents

Unit 6

Adding and Subtracting 10 on Number Grids 111

Math Boxes 6-1 112

Comparing Numbers of Pennies 113

Math Boxes 6-2 114

Math Boxes 6-3 115

Using Helper Doubles Facts 116

Practicing Subtraction Facts 117

Math Boxes 6-4 118

Using Near Doubles to Solve Stories 119

Math Boxes 6-5 120

Math Boxes 6-6 121

My Reference Book Scavenger Hunt 122

Math Boxes 6-7 123

Making Sense of a Problem 124

Math Boxes 6-8: Preview for Unit 7 125

Name-Collection Boxes 126

Finding 10 More and 10 Less 127

Math Boxes 6-9 128

Extending Base-10 Block Riddles 129

Math Boxes 6-10 130

Modeling Number Stories with Equations 131

Math Boxes 6-11 132

Math Boxes 6-12: Preview for Unit 7 133

Unit 7

Fact Families 134

Subtracting 10s 136

Math Boxes 7-1 137

Practicing with Name-Collection Boxes 138

Math Boxes 7-2 139

Fact Families and Fact Triangles 140

Math Boxes 7-3 141

Counting Up and Counting Back 142

Math Boxes 7-4 143

Math Boxes 7-5 144

Solving Garden Number Stories 145

Math Boxes 7-6 146

Math Boxes 7-7 147

Solving "What's My Rule?" Problems 148

Math Boxes 7-8: Preview for Unit 8 149

Function Machines 150

Math Boxes 7-9 151

More "What's My Rule?" Problems 152

Math Boxes 7-10 153

Math Boxes 7-11 154

Math Boxes 7-12: Preview for Unit 8 155

Unit 8

Polygons and Nonpolygons 156

Polygons 157

Fact Families with Fact Triangles and Dominoes 158

Math Boxes 8-1 159

Partitioning Pancakes and Crackers in Halves 160

"What's My Rule?" 161

Math Boxes 8-2 162

Partitioning Crackers and Pancakes in Fourths 163

Math Boxes 8-3 164

Sharing Cheese 165

Math Boxes 8-4 166

Shape Challenge 1 167

Shape Challenge 2 168

Math Boxes 8-5 169

Pictures of 3-Dimensional Shapes 170

Making a Shapes Bar Graph 171

Math Boxes 8-6 172

Building with 3-Dimensional Shapes Record Sheet 173

Sorting by Strategies 174

Drawing and Naming Equal Shares 175

Math Boxes 8-7 176

Math Boxes 8-8: Preview for Unit 9 177

Math Boxes 8-9 178

Number-Grid Puzzles 1 179
Number-Grid Puzzles 2 180
Math Boxes 8-10 181
Solving 10 More, 10 Less Problems 182
Applying and Finding Rules 183
Math Boxes 8-11 184
Math Boxes 8-12: Preview for Unit 9 185

Unit 9

Measuring with Rulers 186
Math Boxes 9-1 187
School Store Mini-Poster 1 188
School Store Mini-Poster 2 189
Recording Number Stories 190
Math Boxes 9-2 191
Buying Toys 192
Math Boxes 9-3 193
Broken-Calculator Puzzles 194
Telling Time 195
Math Boxes 9-4 196
Vending Machine Poster 197
Turning Rectangles into Other Shapes 198
Math Boxes 9-5 199
Vending Machine Prices 200
Creating Vending Machine Number Stories 201
Name-Collection Boxes 202
Math Boxes 9-6 203
Silly Animal Stories 204
Math Boxes 9-7 205
Adding and Comparing Sums 206
Making Smallest and Largest Numbers 207
Math Boxes 9-8 208
Using Different Strategies 209
Math Boxes 9-9 210
Reviewing Attributes of 3-Dimensional Shapes 211
Math Boxes 9-10 212
Partitioning Granola Squares 213
Math Boxes 9-11 214

Math Boxes 9-12 215
My Facts Inventory Record, Part 1 217
My Facts Inventory Record, Part 2 220
My Facts Inventory Record, Part 3 222
My Facts Inventory Record, Part 4 223

Activity Sheets

Clock Face, Hour Hand . . . . Activity Sheet 9
Fact Triangles 1 . . . . Activity Sheet 10
Fact Triangles 2 . . . . Activity Sheet 11
Fact Triangles 3 . . . . Activity Sheet 12
Fact Triangles 4 . . . . Activity Sheet 13
Fact Triangles 5 . . . . Activity Sheet 14
Fact Triangles 6 . . . . Activity Sheet 15
Fact Triangles 7 . . . . Activity Sheet 16

Adding and Subtracting 10 on Number Grids

Use the number grid on the inside back cover of your journal.

1. Close your eyes and put a finger on your number grid.

   Then open your eyes.

   I pointed to _____.

   What number is 10 more than the number? _____

   What number is 10 less than the number? _____

   Do the same thing 4 more times.

2. I pointed to _____.

   What number
   is 10 more? _____

   What number
   is 10 less? _____

3. I pointed to _____.

   What number
   is 10 more? _____

   What number
   is 10 less? _____

4. I pointed to _____.

   What number
   is 10 more? _____

   What number
   is 10 less? _____

5. I pointed to _____.

   What number
   is 10 more? _____

   What number
   is 10 less? _____

Math Boxes

1

Brody's Books	
Fairy Tale	///
Adventure	~~HHT~~ /
Animal	~~HHT~~ //

How many books does Brody have in all?

_____ books

2 Use your feet (with your shoes on) to measure.

About how long is the path from your desk to your teacher's desk?

_____ feet

3 Sofia picked 3 daffodils, 5 daisies, and 2 roses. How many flowers did she pick?

$3 + 5 + 2 =$ □

Unit

4 Add using the number grid.
$71 + 10 = ?$
Choose the best answer.

⬭ 18   ⬭ 72   ⬭ 80   ⬭ 81

Unit

5 **Writing/Reasoning** Solve. Exchange if you need to.

Write <, >, or = to compare.

▊▊▊▊▊▊▯▯▯▯▯ + ▯▯▯▯▯ □ 70

How do you know?

Comparing Numbers of Pennies

Circle who has more pennies.
Use an addition number model to find the difference.

Example:

Carlos   Ⓟ Ⓟ Ⓟ

0 1 2 3 4 5 6 7 8 9

(Lynn)   Ⓟ Ⓟ Ⓟ Ⓟ Ⓟ Ⓟ Ⓟ

Addition Number Model: $\underline{3 + ? = 7}$

How many more pennies? __4__ pennies

1 Amy | 13 pennies |

13 14 15 16 17 18 19 20 21 22

Deon | 21 pennies |

Addition Number Model: _____

How many more pennies? _____ pennies

2 Pete | 7 pennies |

7 8 9 10 11 12 13 14 15 16

Sandy | 15 pennies |

Addition Number Model: _____

How many more pennies? _____ pennies

Math Boxes

Math Boxes

1 Write <, >, or =.

9 ☐ 10

10 ☐ 80

9 ☐ 80

90 ☐ 80

2 It is about 3 o'clock. Draw the hour hand to show the time.

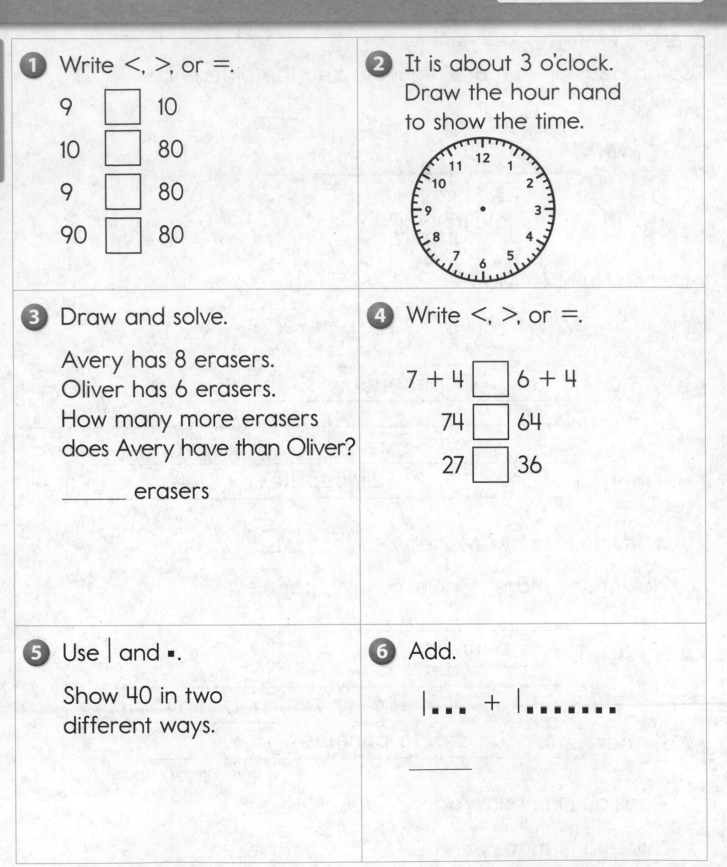

3 Draw and solve.

Avery has 8 erasers. Oliver has 6 erasers. How many more erasers does Avery have than Oliver?

_____ erasers

4 Write <, >, or =.

7 + 4 ☐ 6 + 4

74 ☐ 64

27 ☐ 36

5 Use | and ▪.

Show 40 in two different ways.

6 Add.

|▪▪▪ + |▪▪▪▪▪▪▪

_____

Math Boxes

Math Boxes

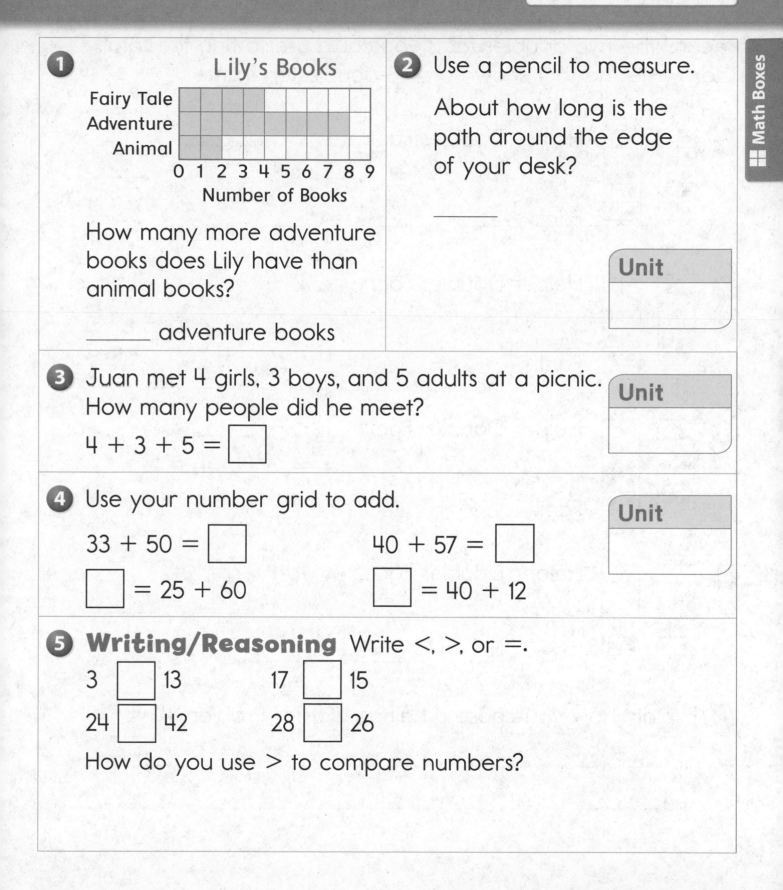

1 Lily's Books

Fairy Tale
Adventure
Animal

0 1 2 3 4 5 6 7 8 9
Number of Books

How many more adventure books does Lily have than animal books?

_____ adventure books

2 Use a pencil to measure.

About how long is the path around the edge of your desk?

_____

Unit

3 Juan met 4 girls, 3 boys, and 5 adults at a picnic. How many people did he meet?

$4 + 3 + 5 = \boxed{}$

Unit

4 Use your number grid to add.

$33 + 50 = \boxed{}$ $40 + 57 = \boxed{}$

$\boxed{} = 25 + 60$ $\boxed{} = 40 + 12$

Unit

5 **Writing/Reasoning** Write <, >, or =.

3 $\boxed{}$ 13   17 $\boxed{}$ 15

24 $\boxed{}$ 42   28 $\boxed{}$ 26

How do you use > to compare numbers?

Using Helper Doubles Facts

Record the two doubles facts you could use to find the total.
Then write the fact shown on the double ten frame.

1 Helper Doubles Fact: _____ + _____ = _____ or

_____ + _____ = _____

Fact: _____ + _____ = _____

2 Helper Doubles Fact: _____ + _____ = _____ or

_____ + _____ = _____

Fact: _____ + _____ = _____

3 Helper Doubles Fact: _____ + _____ = _____ or

_____ + _____ = _____

Fact: _____ + _____ = _____

4 Helper Doubles Fact: _____ + _____ = _____ or

_____ + _____ = _____

Fact: _____ + _____ = _____

5 Explain how you chose a helper fact for Problem 4.

_____

_____

_____

Practicing Subtraction Facts

Solve.

1 _____ $- 5 = 5$ $10 =$ _____ $- 2$ $8 = 16 -$ _____

2 _____ $- 4 = 10$ $10 = 19 -$ _____   _____ $= 14 - 7$

3
$$\begin{array}{r} 18 \\ -\ 3 \\ \hline \end{array}$$
$$\begin{array}{r} 13 \\ -\ 4 \\ \hline \end{array}$$
$$\begin{array}{r} 8 \\ -\ 4 \\ \hline \end{array}$$

4 $17 -$ _____ $= 10$ $12 - 6 =$ _____ $8 -$ _____ $= 8$

5 $10 -$ _____ $= 10$ $2 = 11 -$ _____ $19 -$ _____ $= 2$

Try This

6 $999 - 0 =$ _____

7 Explain how you solved Problem 6.

_____

_____

_____

Math Boxes

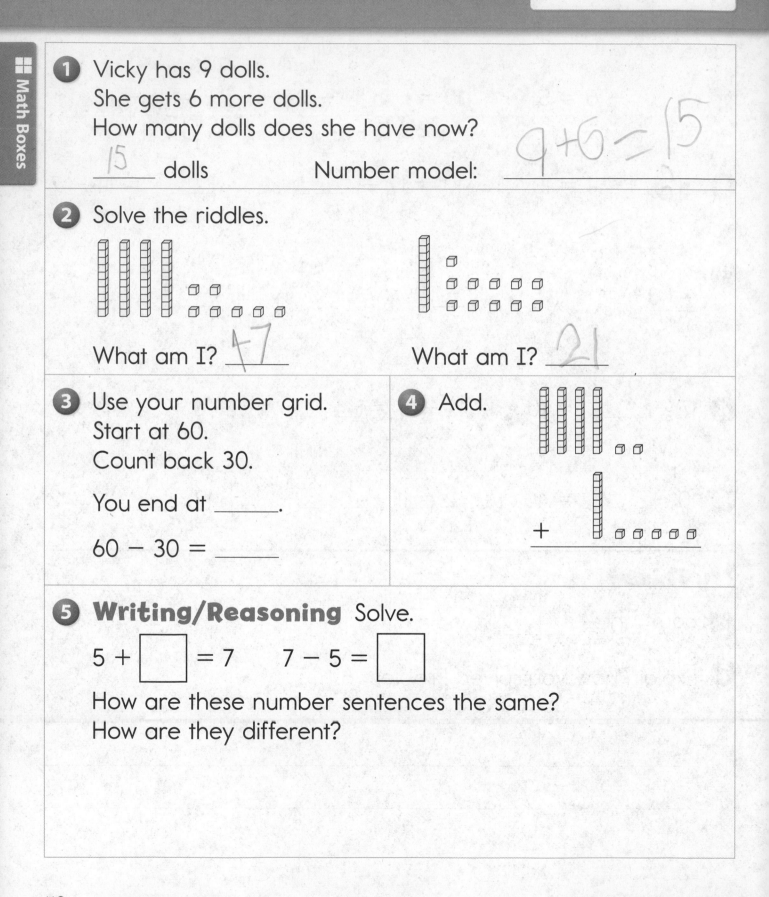

1 Vicky has 9 dolls.
She gets 6 more dolls.
How many dolls does she have now?

___15___ dolls   Number model: ___9+6=15___

2 Solve the riddles.

What am I? __17__   What am I? __21__

3 Use your number grid.
Start at 60.
Count back 30.

You end at _____.

60 − 30 = _____

4 Add.

$+$

5 **Writing/Reasoning** Solve.

5 + ☐ = 7   7 − 5 = ☐

How are these number sentences the same?
How are they different?

Using Near Doubles to Solve Stories

Solve two different ways.
Record your strategies using words and numbers.

I have 5 pencils in my desk and 7 pencils in my backpack.
How many pencils do I have all together?

___12___ pencils

First way

Helper fact: ___6___ + ___6___ = ___12___

Tell how you used the helper fact to find the answer
to the number story.

Second way

Helper fact: ___5___ + ___5___ = ___10___

Tell how you used the helper fact to find the answer
to the number story.

Math Boxes

1 Write <, >, or =.

24 ☒ 30

21 ☒ 12

5 + 5 ☒ 10

16 ☒ 4 + 6

2 Draw the hour hand to show that the time is a little after 8 o'clock.

3 Draw and solve.

Kristen has 3 stars.
Bella has 10 stars.
How many fewer stars does Kristen have than Bella?

_____ fewer stars

4 Circle the larger sum.

11 + 5

12 + 3

Unit
dolls

Write a number sentence using > or < to show which is larger.

_____

5 Use Ⓓ and Ⓟ.

Show 13 cents two different ways.

6 Add.

|····· + |····· = _____

Use | and ▪ to show the sum.

Math Boxes

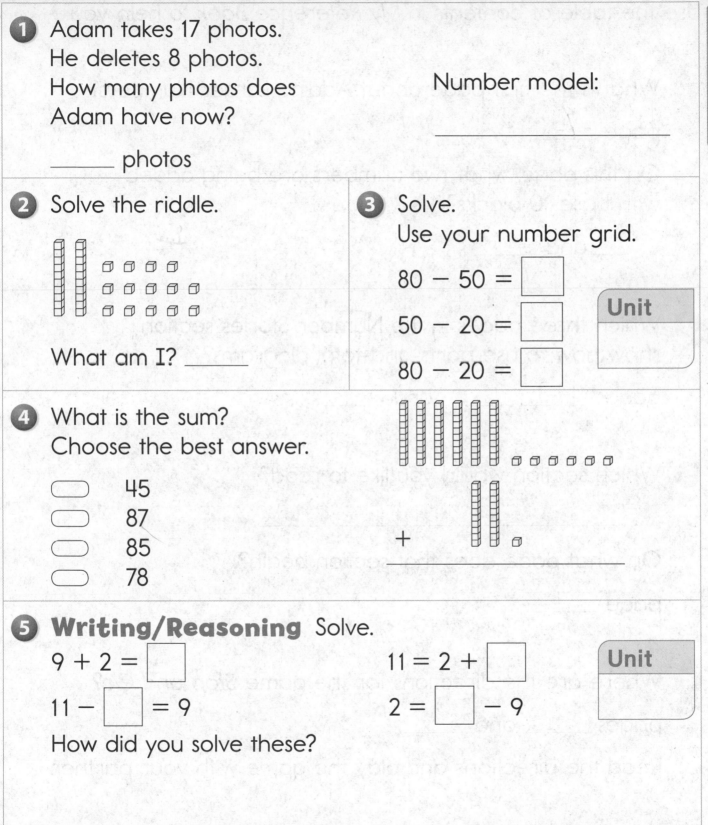

1 Adam takes 17 photos.
He deletes 8 photos.
How many photos does
Adam have now?

Number model:

_____

_____ photos

2 Solve the riddle.

What am I? _____

3 Solve.
Use your number grid.

80 − 50 = ☐

50 − 20 = ☐

80 − 20 = ☐

Unit

4 What is the sum?
Choose the best answer.

⬭ 45
⬭ 87
⬭ 85
⬭ 78

+

5 **Writing/Reasoning** Solve.

9 + 2 = ☐

11 − ☐ = 9

11 = 2 + ☐

2 = ☐ − 9

Unit

How did you solve these?

one hundred twenty-one   121

Use the table of contents in *My Reference Book* to help you.

1 What is the first page about Adding Larger Numbers?

page __76__

On this page, what two numbers are being added with base-10 blocks?

__17__ and __20__

2 Which three pages in the Number Stories section show how to use parts-and-total diagrams?

pages __24__, __25__, and __26__

3 Which section would you like to read?

__26__

On what page does that section begin?

page _____

4 Where are the directions for the game *Stop and Go*?

pages _____ and _____

Read the directions and play the game with your partner.

Math Boxes

1 Solve. Then circle the tens digit.

$6 + 6 =$ _____

MRB 73

2 Which number is 10 less than 65?
Use tools if you like.
Choose the best answer.

◯ 70   ◯ 60
◯ 55   ◯ 5

3 Choose one of these units:
square pattern blocks, crayons, your hands
Use the unit to measure your shoe.
How long is your shoe?
Be sure to write the unit for your answer.

_____

MRB 98

4 What is the number?

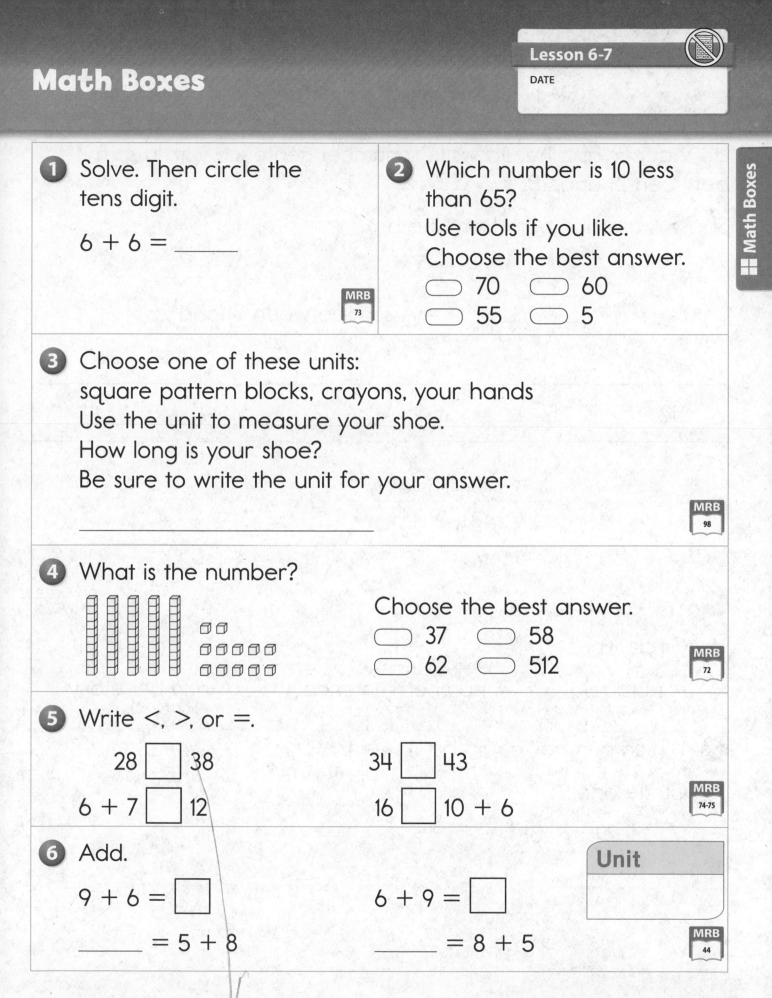

Choose the best answer.

◯ 37   ◯ 58
◯ 62   ◯ 512

MRB 72

5 Write <, >, or =.

28 ☐ 38   34 ☐ 43

6 + 7 ☐ 12   16 ☐ 10 + 6

MRB 74-75

6 Add.

$9 + 6 =$ ☐ $6 + 9 =$ ☐

_____ $= 5 + 8$ _____ $= 8 + 5$

Unit

_____

MRB 44

one hundred twenty-three   123

Making Sense of a Problem

Li's teacher asks her to write a number sentence with a sum between 21 and 25.

She will use numbers from the box.

_____ + _____ = A sum between 21 and 25

12	5	15	17	9	14

1 What should Li write?

Circle one.

A number   A number sentence   A number story

2 Which number could be Li's sum?

Circle one.

17   24   31

Math Boxes
Preview for Unit 7

1 Circle all of the activities that take about 1 minute.

brushing your teeth

baking cookies

watching a movie

singing a song

2 This is a rectangle.

Draw a different rectangle.

MRB
127-128

3 Draw a shape that has 4 sides and 4 vertices.

Use your Pattern-Block Template.

MRB
127

4 Write one number sentence that could go with this domino.

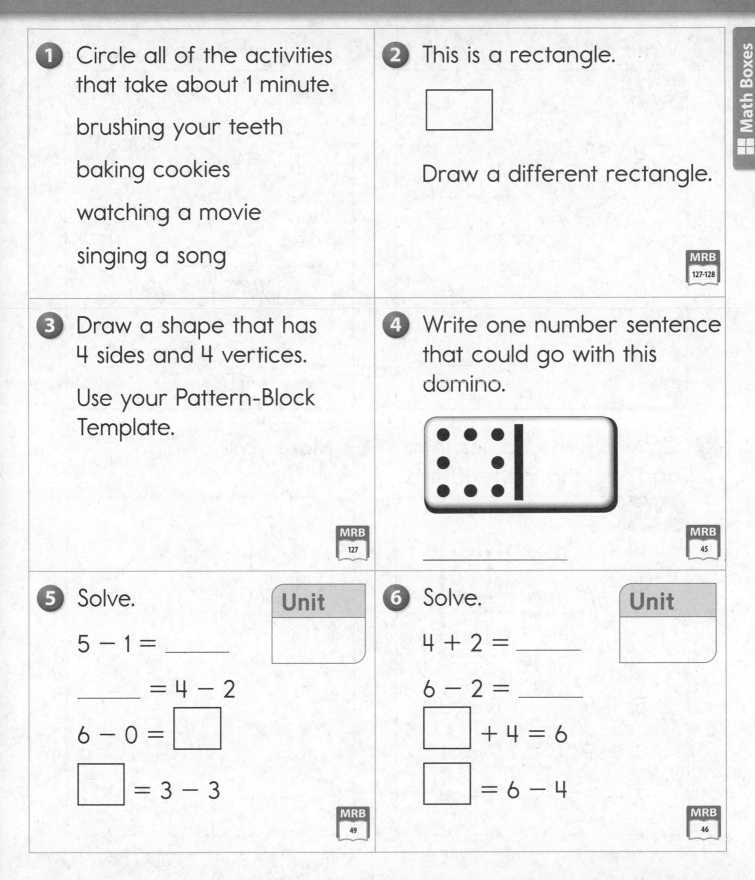

_____

MRB
45

5 Solve.

Unit

$5 - 1 =$ _____

_____ $= 4 - 2$

$6 - 0 =$ ☐

☐ $= 3 - 3$

MRB
49

6 Solve.

Unit

$4 + 2 =$ _____

$6 - 2 =$ _____

☐ $+ 4 = 6$

☐ $= 6 - 4$

MRB
46

Name-Collection Boxes

1 Write other names for 11.

11

8 + 3

13 − 2

2 Write other names for 12.

12

1 dozen

~HHH~ ~HHT~ //

3 + 3 + 3 + 3
15 − 3

3 Cross out the names that don't belong in the 10-box.

10

~HHH~ ~HHT~

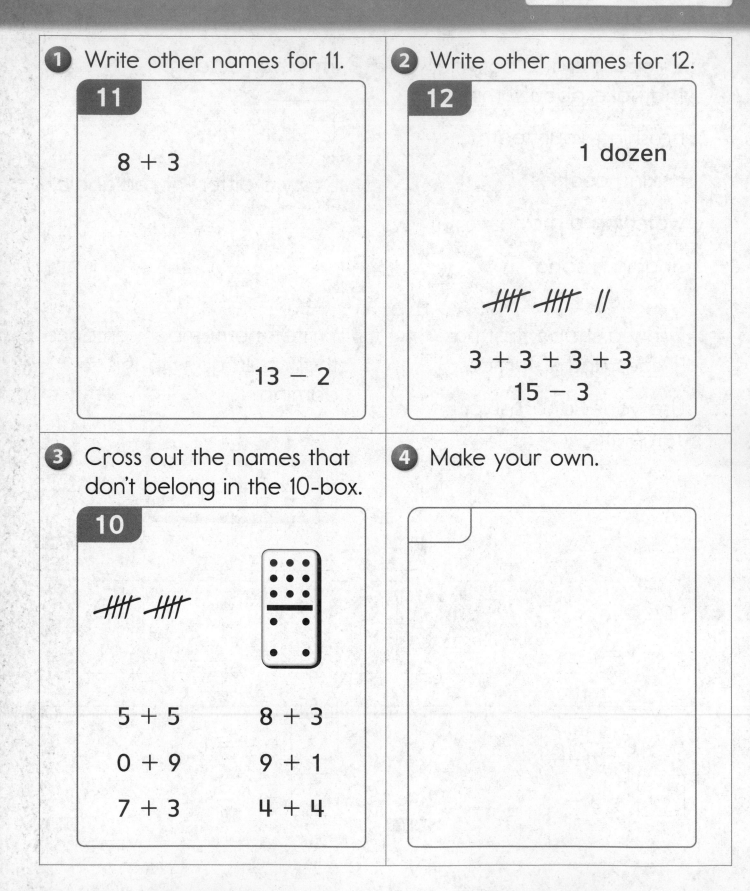

5 + 5   8 + 3

0 + 9   9 + 1

7 + 3   4 + 4

4 Make your own.

Finding 10 More and 10 Less

Use base-10 blocks.

1 Use | and ▪ to show the number 31.

2 Use | and ▪ to show 10 more than 31.

3 Use | and ▪ to show 10 less than 31.

4 Find 76 on your number grid.

What is 10 more than 76? _____

What is 10 less than 76? _____

5 Is it easier for you to use blocks or the number grid to find 10 more and 10 less? Explain.

_____

_____

_____

Math Boxes

1 Count up by 10s.

25, _____, _____,

_____, _____, _____,

_____, _____

2 True or false?

$5 + 6 + 4 = 5 + 10$

_____

Unit

MRB
52

3 What is the tens digit in 78?

_____

What is the tens digit in 87?

_____

MRB
73

4 Add.

$7 + 3 + 2 =$ _____

$7 + 5 =$ _____

$10 + 2 =$ _____

Unit

MRB
44

5 **Writing/Reasoning**

Betty has 25 red pens and 35 blue pens.
Archie has 35 red pens and 25 blue pens.
Do Betty and Archie have the same number of pens?
How do you know?

MRB
44

Extending Base-10 Block Riddles

Hundreds	Tens	Ones

Solve the riddles. Use base-10 blocks to help you.

1 What am I? _____

2 What am I? _____

3 1 long and 8 cubes

What am I? _____

How many tens? _____

4 4 longs and 0 cubes

What am I? _____

How many ones? _____

5 2 longs and 14 cubes

What am I? _____

How many tens? _____

6 4 longs and 21 cubes

What am I? _____

How many ones? _____

Try This

7 1 flat, 4 longs, and 4 cubes. What am I? _____

8 2 flats, 7 longs, and 12 cubes. What am I? _____

9 Use | and ▪. Show a number with the same tens digit as 122.

Math Boxes

Math Boxes

1 Solve. Then circle the ones digit.

8 + 8 = _____

MRB
73

2 What number is 10 more than 90?

Use tools if you like.

_____

3 Choose one:

your shoe
your pencil
an eraser

Use it to measure the path from your desk to the door.

How long is the path?

Be sure to write the unit.

_____

MRB
98

4 Solve the riddles.

What am I?   What am I?

_____   _____

MRB
72

5 Write <, >, or =.

6 ☐ 8

16 ☐ 8

16 ☐ 81

1 + 60 ☐ 18

MRB
74-75

6 Solve.

☐ = 4 + 7

7 + ☐ = 11

☐ + 4 = 14

14 = 10 + ☐

Unit

Modeling Number Stories with Equations

Solve the number story. Then draw a line from the number story to the equation that matches it.

1 Janine ate 7 berries.
Carlos ate 10 berries.
How many more berries did
Carlos eat than Janine?

_____ more berries

$10 - 7 = \boxed{}$

$3 + \boxed{} = 8$

2 Shavon found 3 leaves.
Janelle found 8 leaves.
How many fewer leaves did
Shavon find than Janelle?

_____ fewer leaves

$8 - 3 = \boxed{}$

$10 + \boxed{} = 19$

3 A cat weighs 10 pounds.
A dog weighs 19 pounds.
How much more does the
dog weigh than the cat?

_____ more pounds

$19 - 10 = \boxed{}$

$7 + \boxed{} = 10$

Math Boxes

Math Boxes

① Count back by 10s.

99, _____, _____,

_____, _____, _____,

_____, _____

② Circle all of the names for 10.

4 + 8   15 − 5

17 − 8   7 + 3

~~HHI~~ ~~HHI~~ //

MRB 53

③ What is the tens digit in 50?

_____

What is the tens digit in 56?

_____

MRB 73

④ Solve.

8 + 4 + 2 = _____

8 + 6 = _____

12 + 2 = _____

MRB 44

⑤ **Writing/Reasoning** Write <, >, or =.

10 + 7 ☐ 20

7 + 10 ☐ 71

How does filling in the first box help you fill in the second box?

Unit

MRB 44, 75

132   one hundred thirty-two

Math Boxes
Preview for Unit 7

1 Name an activity that takes about 1 minute.

2 Fill in the rule and the frames.

Rule

| | 43 | 33 | | |

MRB
55

3 Draw a shape with 3 straight sides.

What is your shape called?

_____

MRB
122

4 Write a subtraction number sentence that could go with this domino.

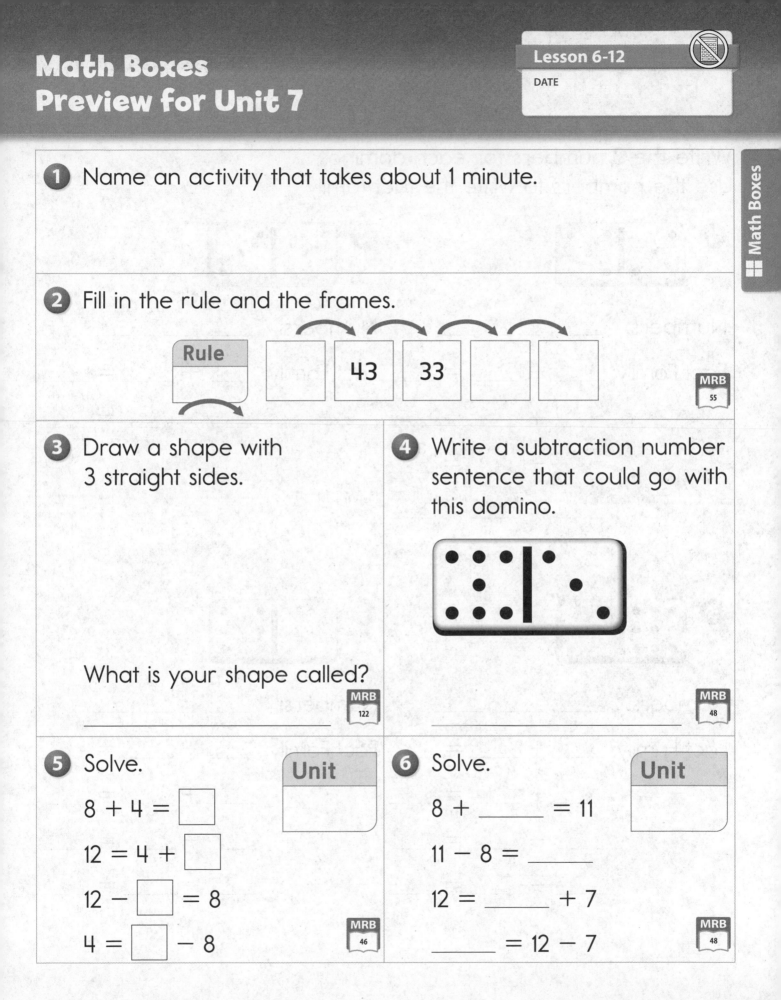

_____

MRB
48

5 Solve.

Unit

$8 + 4 =$ ☐

$12 = 4 +$ ☐

$12 -$ ☐ $= 8$

$4 =$ ☐ $- 8$

MRB
46

6 Solve.

Unit

$8 +$ _____ $= 11$

$11 - 8 =$ _____

$12 =$ _____ $+ 7$

_____ $= 12 - 7$

MRB
48

Fact Families

Write the 3 numbers for each domino.
Use the numbers to write the fact family.

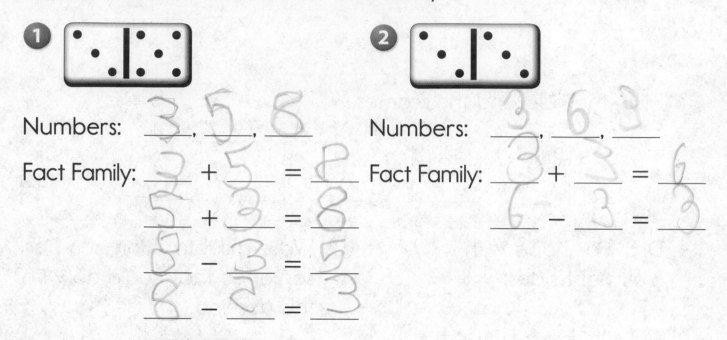

1 Numbers: _3_, _5_, _8_

Fact Family: _3_ + _5_ = _8_

_5_ + _3_ = _8_

_8_ − _3_ = _5_

_8_ − _5_ = _3_

2 Numbers: _3_, _6_, _3_

Fact Family: _3_ + _3_ = _6_

_6_ − _3_ = _3_

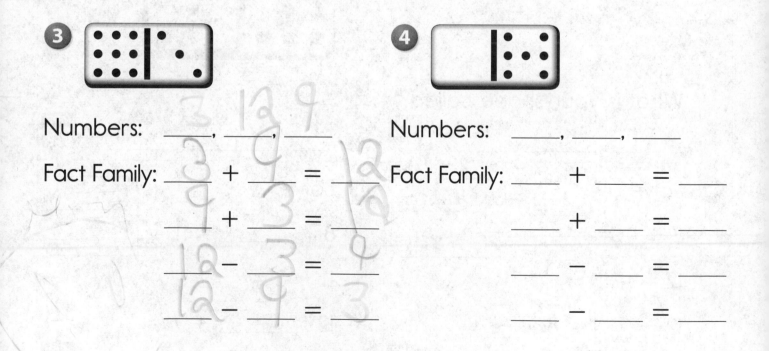

3 Numbers: _3_, _9_, _9_

Fact Family: _3_ + _9_ = _12_

_9_ + _3_ = _12_

_12_ − _3_ = _9_

_12_ − _9_ = _3_

4 Numbers: ___, ___, ___

Fact Family: ___ + ___ = ___

___ + ___ = ___

___ − ___ = ___

___ − ___ = ___

5 Abbey covers one side of a domino.
You can only see 4 dots.
She says that the total for the domino is 9.
Complete the fact family.

Numbers: 4, __5__, 9

Fact Family: 4 + ____ = 9

____ + ____ = ____

____ − ____ = ____

____ − ____ = ____

6 Juan does not know the answer to 12 − 6 = ☐.
What addition fact could he use to help him find the answer?
Explain.

_____

_____

_____

Subtracting 10s

Solve.

1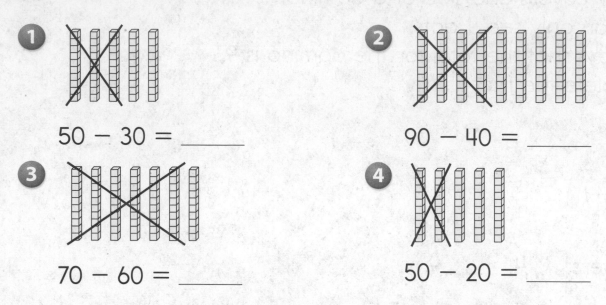

50 − 30 = _____

2

90 − 40 = _____

3

70 − 60 = _____

4

50 − 20 = _____

Draw an X to show the subtraction problem.
Then solve.

5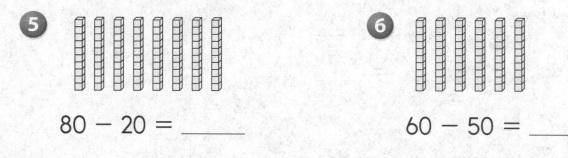

80 − 20 = _____

6

60 − 50 = _____

Try This

Write a number sentence for the picture. Solve.

7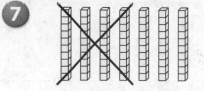

Number sentence: _____

Math Boxes

Math Boxes

1 Record the time.

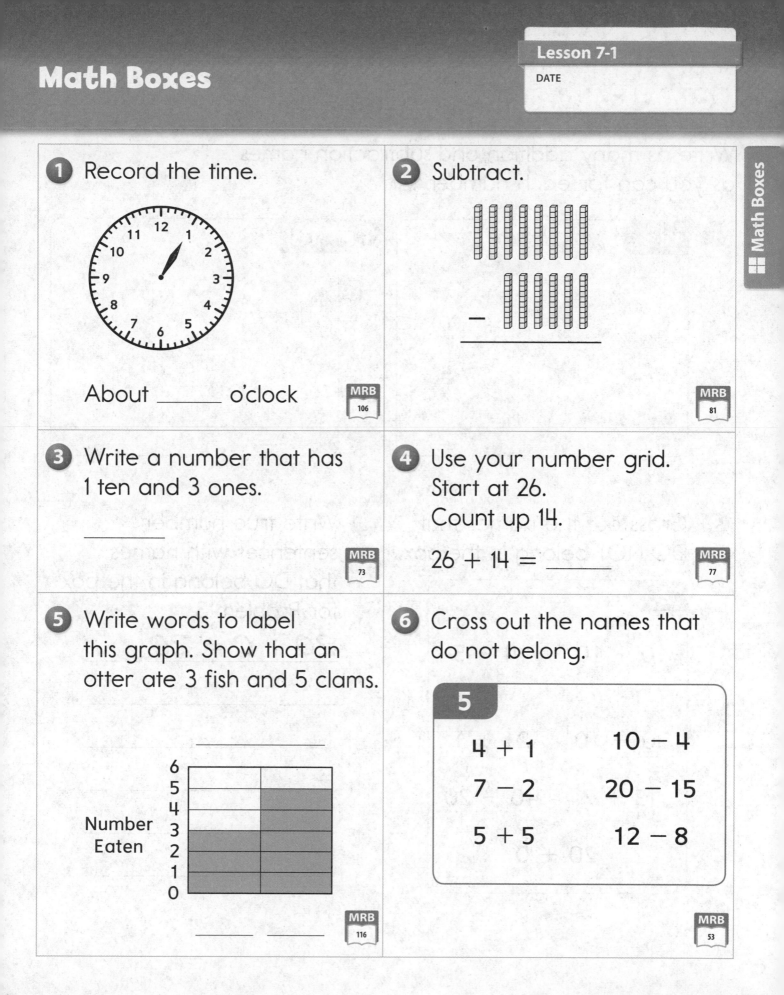

About _____ o'clock

MRB
106

2 Subtract.

MRB
81

3 Write a number that has 1 ten and 3 ones.

_____

MRB
73

4 Use your number grid.
Start at 26.
Count up 14.

$26 + 14 =$ _____

MRB
77

5 Write words to label this graph. Show that an otter ate 3 fish and 5 clams.

_____

Number Eaten

6
5
4
3
2
1
0

_____   _____

MRB
116

6 Cross out the names that do not belong.

5
$4 + 1$ $10 - 4$
$7 - 2$ $20 - 15$
$5 + 5$ $12 - 8$

MRB
53

one hundred thirty-seven   137

Practicing with Name-Collection Boxes

Write as many addition and subtraction names
as you can for each number.

1 | **13**

2 | **18**

3 Cross out the names that
DO NOT belong in the box.

20

10 + 10	14 + 5
30 − 10	21 + 1
13 + 7	40 − 20
20 + 0	

4 Write true number
sentences with names
that DO belong in the box
for Problem 3.

20 + 0 = 20

_____

_____

_____

_____

Math Boxes

1 Use a marker to measure.

How tall is your desk or table?

_____ markers

MRB
98

2 Write the missing numbers.

Rule
Add 10

| 50 | | 70 | | 90 | |

3 What is the value of the 3 in 38?

○ 3
○ 8
○ 10
○ 30

MRB
73

4 Write 3 names.

17

MRB
53

5 **Writing/Reasoning** Solve. 3 + 5 + 2 = _____

Which numbers did you add first? Why did you choose those?

MRB
44

Math Boxes

one hundred thirty-nine   139

Fact Families and Fact Triangles

Write the fact family for each Fact Triangle.

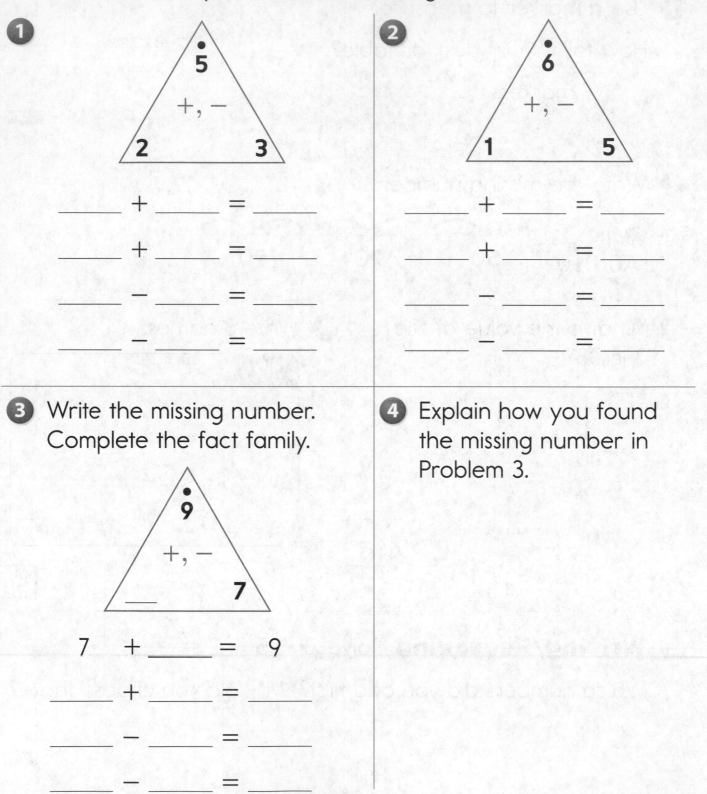

1

$\bullet$5

+, −

2   3

____ + ____ = ____

____ + ____ = ____

____ − ____ = ____

____ − ____ = ____

2

$\bullet$6

+, −

1   5

____ + ____ = ____

____ + ____ = ____

____ − ____ = ____

____ − ____ = ____

3 Write the missing number. Complete the fact family.

$\bullet$9

+, −

____   7

7 + ____ = 9

____ + ____ = ____

____ − ____ = ____

____ − ____ = ____

4 Explain how you found the missing number in Problem 3.

Math Boxes

Math Boxes

1 Record the time.

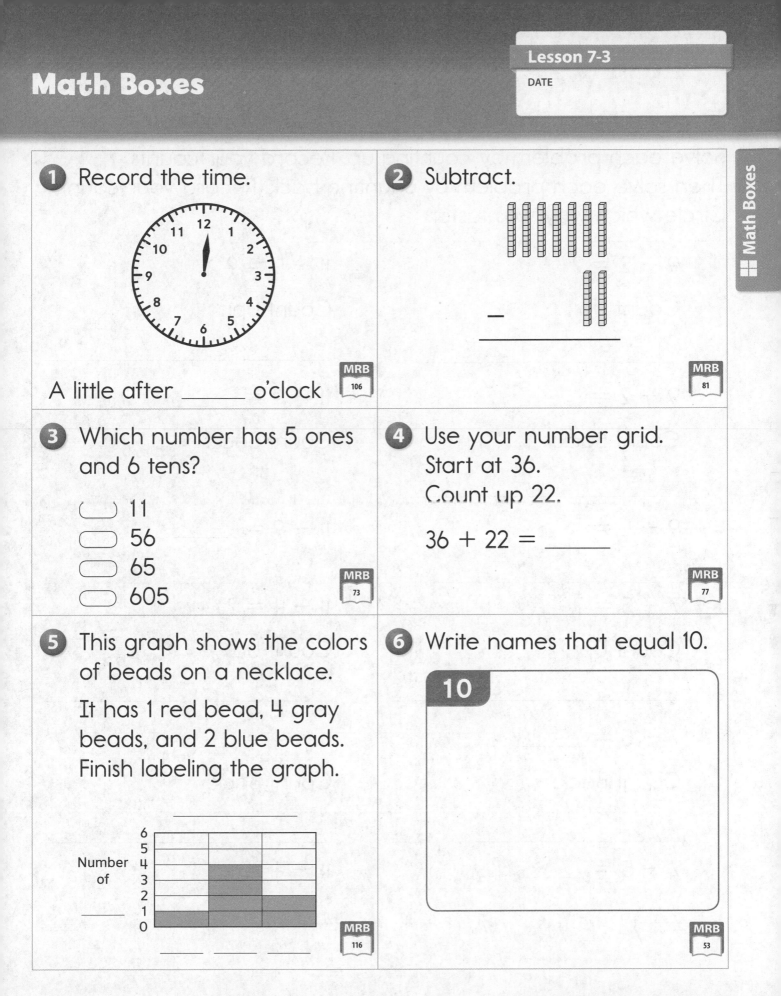

A little after _____ o'clock

MRB 106

2 Subtract.

MRB 81

3 Which number has 5 ones and 6 tens?

- ⬭ 11
- ⬭ 56
- ⬭ 65
- ⬭ 605

MRB 73

4 Use your number grid.
Start at 36.
Count up 22.

$36 + 22 =$ _____

MRB 77

5 This graph shows the colors of beads on a necklace.

It has 1 red bead, 4 gray beads, and 2 blue beads. Finish labeling the graph.

Number of _____

6
5
4
3
2
1
0

MRB 116

6 Write names that equal 10.

10

MRB 53

Counting Up and Counting Back

Solve each problem by counting up. Record your counts.
Then solve each problem by counting back. Record your counts.
Circle which way was faster.

1 $9 - 2 = ?$

Count Up:

_____

$9 - 2 =$ _____

Count Back:

_____

$9 - 2 =$ _____

2 $11 - 9 = ?$

Count Up:

_____

$11 - 9 =$ _____

Count Back:

_____

$11 - 9 =$ _____

3 $7 - 5 = ?$

Count Up:

_____

$7 - 5 =$ _____

Count Back:

_____

$7 - 5 =$ _____

4 $11 - 4 = ?$

Count Up:

_____

$11 - 4 =$ _____

Count Back:

_____

$11 - 4 =$ _____

Math Boxes

1 Subtract.

8 − 4 = _____

12 − 6 = _____

18 − 9 = _____

14 − 7 = _____

MRB 48-49

2 Use your number grid. Write <, >, or =.

3 + 5 ☐ 30 + 50

30 + 5 ☐ 3 + 50

70 − 40 ☐ 7 − 4

74 ☐ 70 + 4

MRB 75

3 Find the sums. Use a number grid if you like.

Unit

2 + 5 = _____

12 + 5 = _____

42 + 5 = _____

102 + 5 = _____

MRB 77

4 Jess writes 8 stories about horses and 6 stories about people.

Unit

How many stories does she write in all? _____

Number model: _____

MRB 24

5 Stan and Ian each swim 4 laps. Cassidy swims 5 laps.

How many laps do they swim in all?

4 + 4 + 5 = ☐

Unit

MRB 44

6 Subtract.   4 − 2 = _____

What addition fact can you use to solve 4 − 2?

_____

MRB 48

143

Math Boxes

1 Use a marker to measure.

How tall is your teacher's desk?

_____ markers

MRB
98

2 Add or subtract 10.

50 _____

70 _____

90 _____

3 What is the value of the 9 in 89?

_____

MRB
73

4 Label the name-collection box.

$14 + 6$ $7 + 13$
$5 + 5 + 5 + 5$

MRB
53

5 **Writing/Reasoning** Solve.

$7 + 5 + 4 =$ _____ $4 + 7 + 5 =$ _____

$5 + 4 + 7 =$ _____

Are your answers the same? Explain.

MRB
44

Solve. Then write a number model.

1. Frank grew 7 red flowers.
   He grew 5 more flowers.
   How many flowers did he grow in all? _____ flowers

   Number model: _____

2. Clark planted 18 seeds.
   Ashley planted 15 seeds.
   Who planted more seeds? _____
   How many more? _____ seeds

   Number model: _____

3. In the garden, there are 6 green peppers,
   7 yellow squash, and 4 yellow peppers.
   How many vegetables are in the garden? _____ vegetables

   Number model: _____

Try This

4. For Problem 3, Dan first found the number of peppers.
   Then he added the number of squash.
   Pam first found the number of yellow vegetables.
   Then she added the number of green vegetables.
   Do both ways get the same answer? Which is faster for you?

   _____

   _____

Math Boxes

1 Subtract.

_____ − 8 = 8

Choose the best answer.

⬭ 16   ⬭ 8

⬭ 0   ⬭ 88

MRB
48

2 Use your number grid.
Write <, >, or =.

10 + 23 ☐ 40

18 + 5 ☐ 5 + 18

4 + 26 ☐ 26 + 40

MRB
75, 77

3 Find the sums. Use a
number grid if you like.

16 + 9 = _____

26 + 9 = _____

56 + 9 = _____

106 + 9 = _____

MRB
77

4 Ji plays 7 games with Sandy.
He also plays with Caleb.
Ji plays 15 games in all.
How many games does Ji
play with Caleb?

_____ games

Number model:

_____

MRB
29

5 Eva catches 4 trout,
3 salmon, and 8 bass.

How many fish does
she catch?

4 + 3 + 8 = ☐ fish

MRB
44

6 Subtract.

4 − 1 = ___   7 − 2 = ___

5 − 4 = ___   6 − 0 = ___

What addition fact can you
use to solve 5 − 4 ?

MRB
48-49

Math Boxes

1 Circle the digit in the ones place.

50

What does the 5 mean?

_____

MRB 73

2 What am I?

_____

Use | and ▪ to show this number another way.

MRB 72

3 There are 8 cups.
There are 5 plates.
How many more cups are there than plates?

_____ cups
Number model:

_____

MRB 30

4 Think addition to subtract.

$11 - 8 =$ ____

____ $= 14 - 8$

____ $= 12 - 6$

$10 - 7 =$ ____

MRB 48

5 Writing/Reasoning Label the name-collection box. How do you know what to label the name-collection box?

$3 + 10 + 2$

$20 - 5$

$7 + 8$

MRB 53

Solving "What's My Rule?" Problems

Find the rules and missing numbers.

1

in ↓

Rule
×2

out ↓

in	out
7	5
11	9
4	2
9	7

Your Turn | 12 | 10 |

2

in ↓

Rule
+4

out ↓

in	out
1	5
3	7
4	8
2	6

Your Turn | 7 | 3 |

3

in ↓

Rule

out ↓

in	out
33	43
12	22
50	60
28	

Your Turn | | |

Try This

4

in ↓

Rule

out ↓

in	out
1	2
2	4
3	6
4	
6	

Your Turn | | |

Math Boxes
Preview for Unit 8

Math Boxes

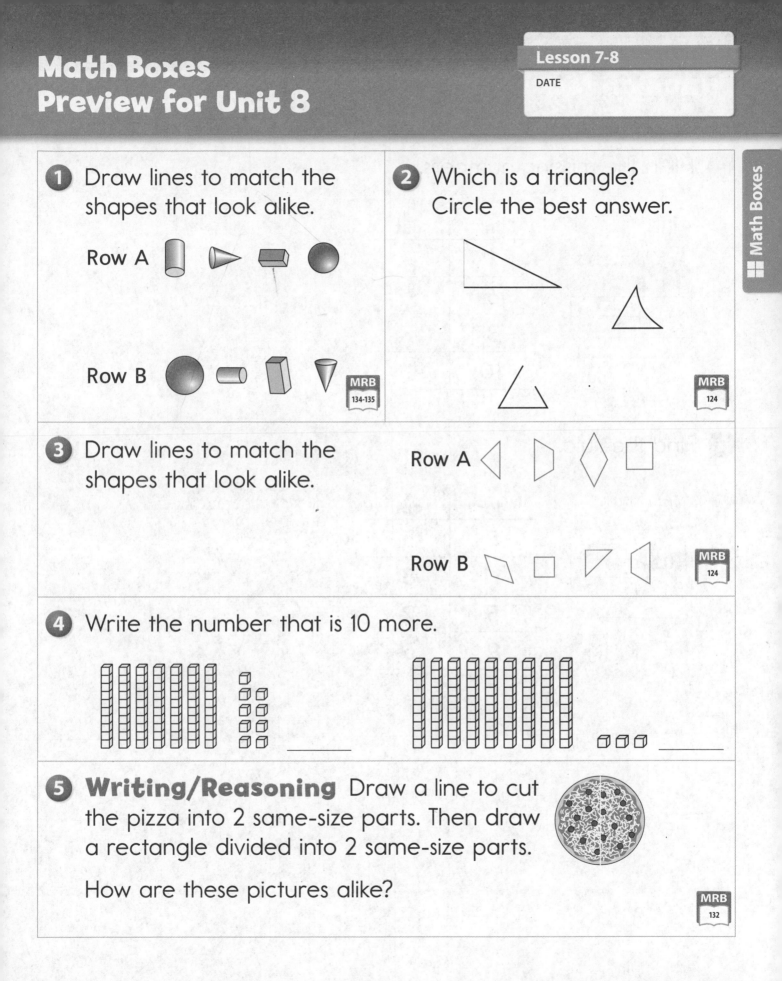

1. Draw lines to match the shapes that look alike.

   Row A

   Row B

   MRB 134-135

2. Which is a triangle? Circle the best answer.

   MRB 124

3. Draw lines to match the shapes that look alike.

   Row A

   Row B

   MRB 124

4. Write the number that is 10 more.

   _____

   _____

5. **Writing/Reasoning** Draw a line to cut the pizza into 2 same-size parts. Then draw a rectangle divided into 2 same-size parts.

   How are these pictures alike?

   MRB 132

Function Machines

1 Fill in the missing numbers.

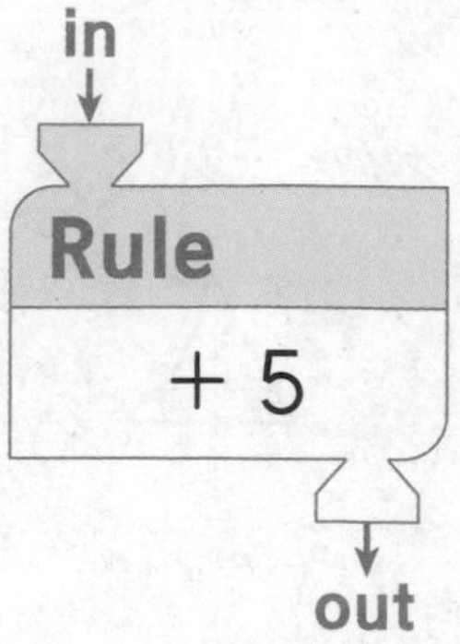

in

Rule
+ 5

out

in	out
1	6
6	11
3	8
10	15

2 Find the rule.

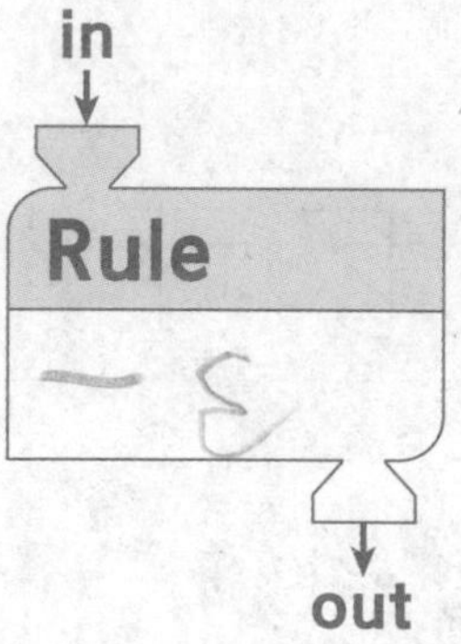

in

Rule
− 5

out

in	out
8	5
11	8
5	2
9	6

Math Boxes

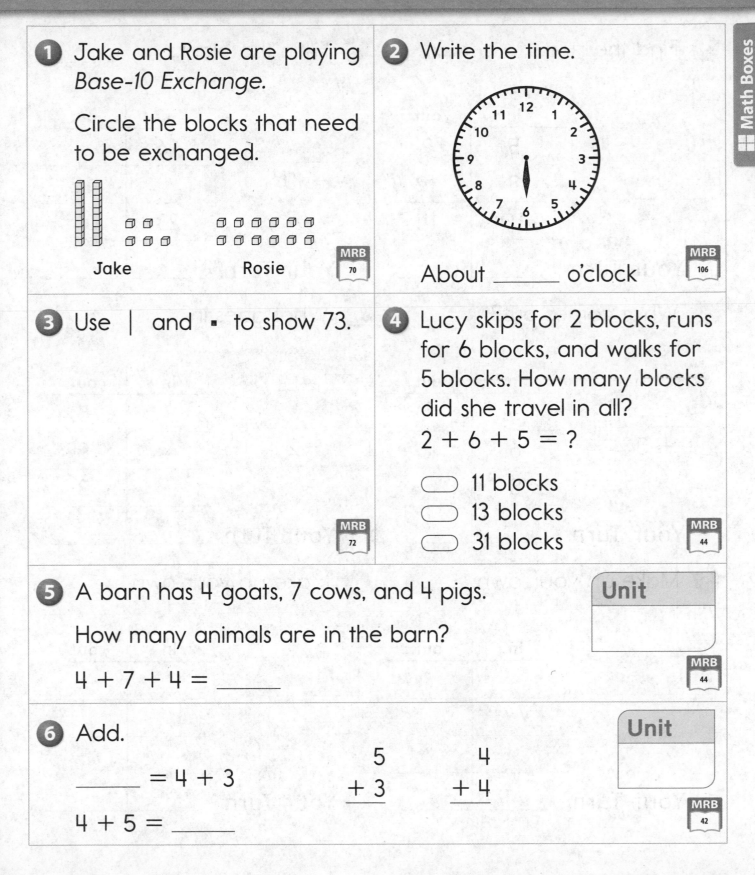

① Jake and Rosie are playing *Base-10 Exchange*.

Circle the blocks that need to be exchanged.

Jake   Rosie

MRB 70

② Write the time.

About _____ o'clock

MRB 106

③ Use | and ▪ to show 73.

MRB 72

④ Lucy skips for 2 blocks, runs for 6 blocks, and walks for 5 blocks. How many blocks did she travel in all?

$2 + 6 + 5 = ?$

○ 11 blocks
○ 13 blocks
○ 31 blocks

MRB 44

⑤ A barn has 4 goats, 7 cows, and 4 pigs.

How many animals are in the barn?

$4 + 7 + 4 =$ _____

Unit

MRB 44

⑥ Add.

_____ $= 4 + 3$

$4 + 5 =$ _____

$\begin{array}{r} 5 \\ + 3 \\ \hline \end{array}$

$\begin{array}{r} 4 \\ + 4 \\ \hline \end{array}$

Unit

MRB 42

More "What's My Rule?" Problems

1 Find the rule.

in ↓

Rule

out ↓

in	out
5	9
8	12
10	14

Your Turn | |

2 Fill in the blanks.

in ↓

Rule
− 10

out ↓

in	out
16	
	12
23	

Your Turn | |

3 What comes out?

in ↓

Rule
+ 5

out ↓

in	out
5	
6	
8	

Your Turn | |

4 What goes in?

in ↓

Rule
− 2

out ↓

in	out
	5
	9
	3

Your Turn | |

5 Make up your own.

in ↓

Rule

out ↓

in	out

Your Turn | |

6 Make up your own.

in ↓

Rule

out ↓

in	out

Your Turn | |

Math Boxes

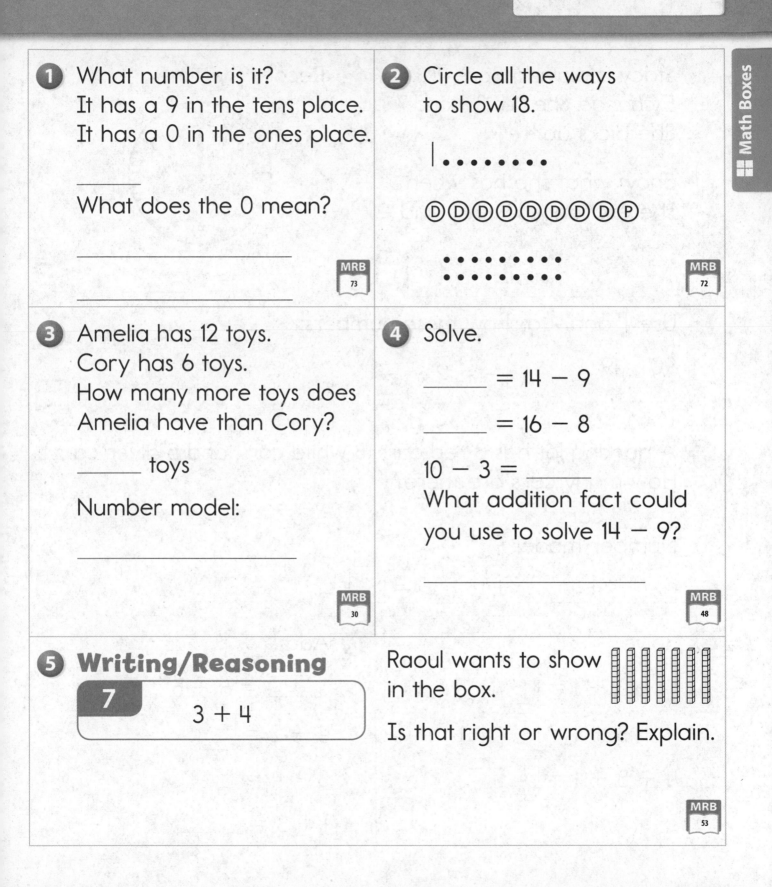

1 What number is it?
It has a 9 in the tens place.
It has a 0 in the ones place.

_____

What does the 0 mean?

_____

_____

MRB 73

2 Circle all the ways
to show 18.

| ●●●●●●●●

ⒹⒹⒹⒹⒹⒹⒹⓅ

●●●●●●●●
●●●●●●●●●

MRB 72

3 Amelia has 12 toys.
Cory has 6 toys.
How many more toys does
Amelia have than Cory?

_____ toys

Number model:

_____

MRB 30

4 Solve.

_____ $= 14 - 9$

_____ $= 16 - 8$

$10 - 3 =$ _____
What addition fact could
you use to solve $14 - 9$?

_____

MRB 48

5 **Writing/Reasoning**

┌─────────────────────┐
│ **7** │
│ $3 + 4$ │
└─────────────────────┘

Raoul wants to show ▯▯▯▯▯▯▯
in the box.

Is that right or wrong? Explain.

MRB 53

Math Boxes

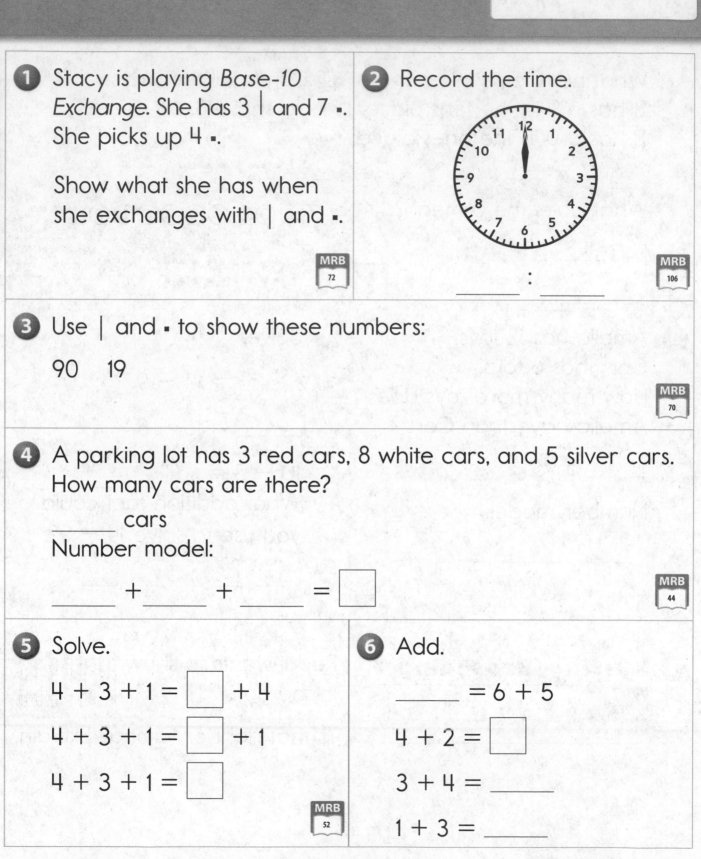

1 Stacy is playing *Base-10 Exchange.* She has 3 | and 7 ▪. She picks up 4 ▪.

Show what she has when she exchanges with | and ▪.

MRB 72

2 Record the time.

_____ : _____

MRB 106

3 Use | and ▪ to show these numbers:

90   19

MRB 70

4 A parking lot has 3 red cars, 8 white cars, and 5 silver cars. How many cars are there?

_____ cars

Number model:

_____ + _____ + _____ = ☐

MRB 44

5 Solve.

$4 + 3 + 1 = \boxed{} + 4$

$4 + 3 + 1 = \boxed{} + 1$

$4 + 3 + 1 = \boxed{}$

MRB 52

6 Add.

$\underline{} = 6 + 5$

$4 + 2 = \boxed{}$

$3 + 4 = \underline{}$

$1 + 3 = \underline{}$

Math Boxes
Preview for Unit 8

Math Boxes

1 How many corners?

_____ corners

MRB 136

2 Draw a polygon with 3 sides.

What shape is it?

_____

MRB 124-125

3 Draw lines to match the shapes that look alike.

Column A   Column B

MRB 122-123

4 Write the numbers that are 10 more and 10 less than each number below. Use base-10 blocks if you need to.

_____ 14 _____

_____ 58 _____

_____ 90 _____

5 **Writing/Reasoning**

How many of the smallest triangles are in this square?

_____ triangles

How do you know if they are the same size?

MRB 133

Polygons and Nonpolygons

These are polygons.

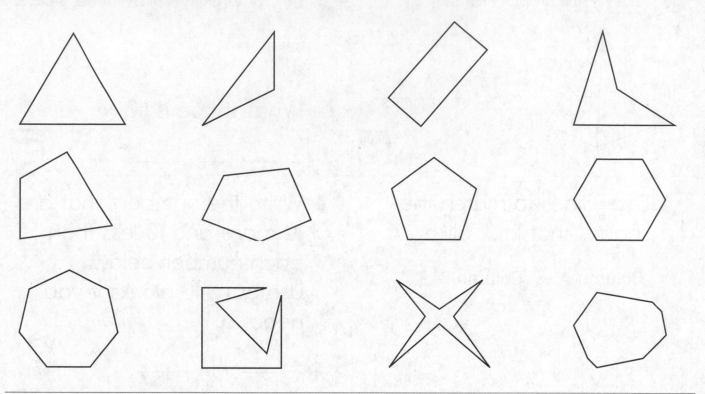

These are not polygons.

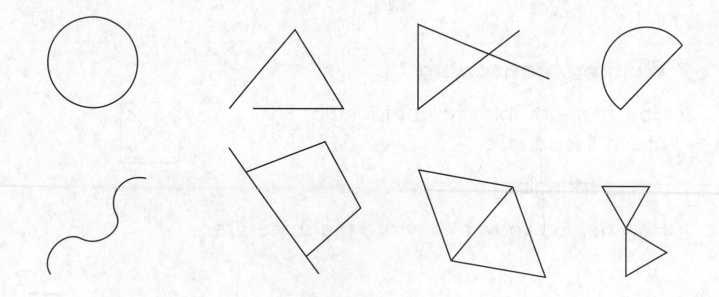

Polygons

Triangles

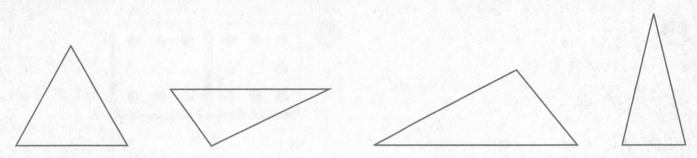

Quadrilaterals

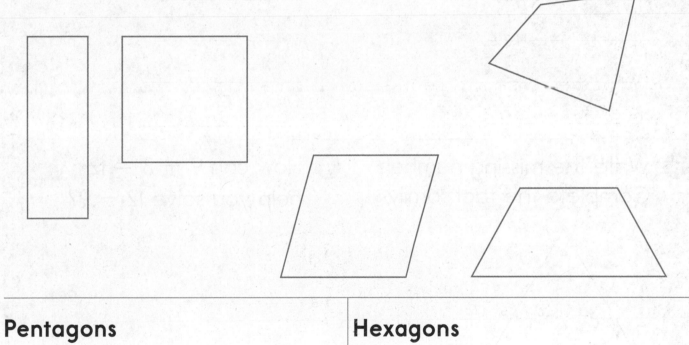

Pentagons

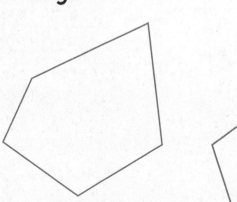

Hexagons

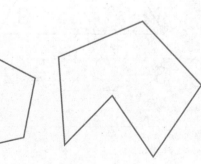

Fact Families with Fact Triangles and Dominoes

Write the fact families.

1

$$\triangle \quad 11 \quad +, - \quad 6 \quad 5$$

_____

_____

_____

_____

2

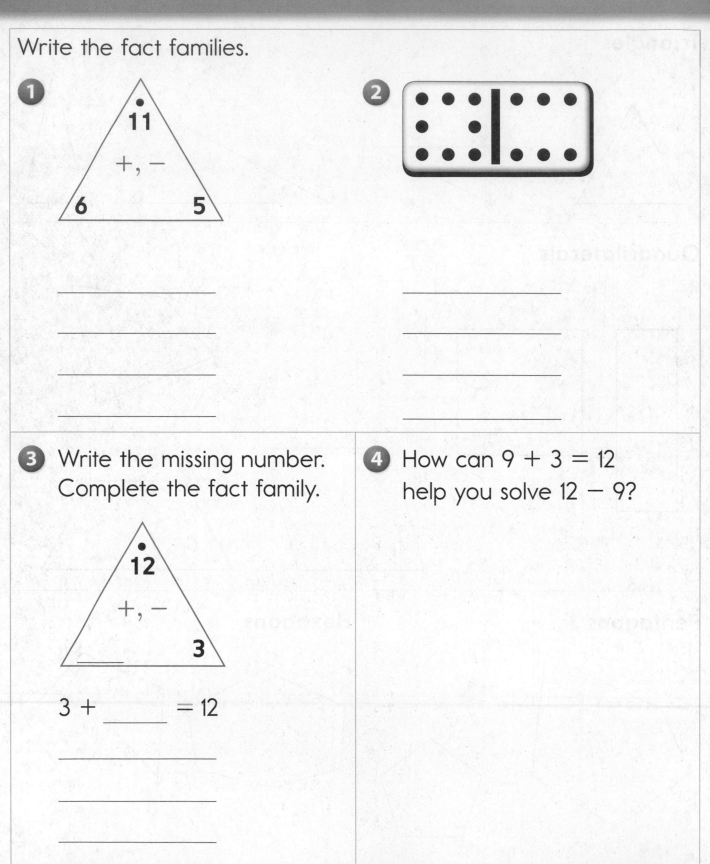

_____

_____

_____

_____

3 Write the missing number. Complete the fact family.

$$\triangle \quad 12 \quad +, - \quad \underline{} \quad 3$$

3 + _____ = 12

_____

_____

_____

4 How can 9 + 3 = 12 help you solve 12 − 9?

Math Boxes

1 Draw a polygon with
4 sides.

How many corners does
it have?

_____ corners

MRB
127

2 Name a classmate.

_____

Choose a unit: pencils,
paperclips, or journals.
Measure your
classmate's arm.
Be sure to write the unit.

_____

MRB
98

3 Solve.
A spider has 8 legs. A dog has 4 legs.
How many more legs does a spider have than a dog?
Choose the best answer.

◯ 4 legs   ◯ 8 legs   ◯ 12 legs   ◯ 16 legs

MRB
30

4 Solve.   18   12   15   10
              − 9   − 8   − 7   − 7

Write an addition fact you could use to solve 10 − 7.

MRB
48

5 **Writing/Reasoning** What am I? |.... _____
Use | and ▪ to show the number another way.
How do you know that they both show the same number?

MRB
72

Partitioning Pancakes and Crackers in Halves

1 Show how you divided the pancake into 2 equal shares.

2 Show how to share 1 cracker between 2 people.

3 Show another way to share 1 cracker between 2 people.

4 Write a name for 1 of the equal parts of the cracker or pancake.

"What's My Rule?"

Find the rule and the missing numbers.
Use the number grid if you like.

1

in	out
14	15
22	23
4	
	19

2

in	out
11	19
24	32
10	
	37

3

in	out
13	3
51	41
	8
	29

4

in	out
23	33
15	25
7	
	37

5 How can you find the rule in Problem 4?

_____

_____

_____

_____

Math Boxes

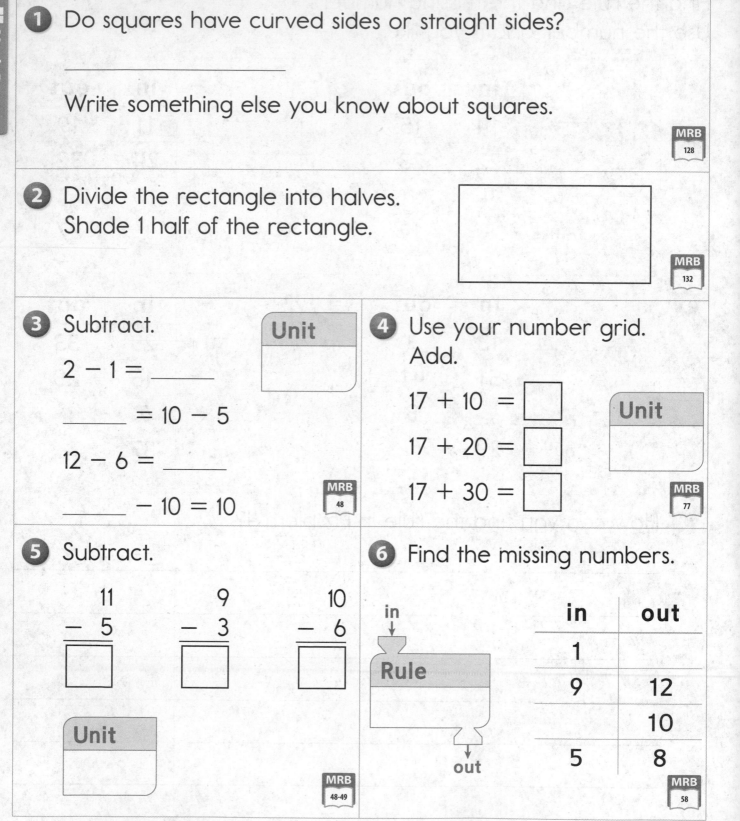

1 Do squares have curved sides or straight sides?

_____

Write something else you know about squares.

MRB 128

2 Divide the rectangle into halves.
Shade 1 half of the rectangle.

MRB 132

3 Subtract.

Unit

$2 - 1 =$ _____

_____ $= 10 - 5$

$12 - 6 =$ _____

_____ $- 10 = 10$

MRB 48

4 Use your number grid.
Add.

$17 + 10 =$ ☐

$17 + 20 =$ ☐

$17 + 30 =$ ☐

Unit

MRB 77

5 Subtract.

$\begin{array}{r} 11 \\ -\ 5 \\ \hline \end{array}$ $\begin{array}{r} 9 \\ -\ 3 \\ \hline \end{array}$ $\begin{array}{r} 10 \\ -\ 6 \\ \hline \end{array}$

Unit

MRB 48-49

6 Find the missing numbers.

in ↓

Rule

out ↓

in	out
1	
9	12
	10
5	8

MRB 58

Partitioning Crackers and Pancakes in Fourths

1 Show how to share 1 cracker equally among 4 people.

2 Show another way to share 1 cracker equally among 4 people.

3 Draw lines to show how to share the pancake equally among 4 people.

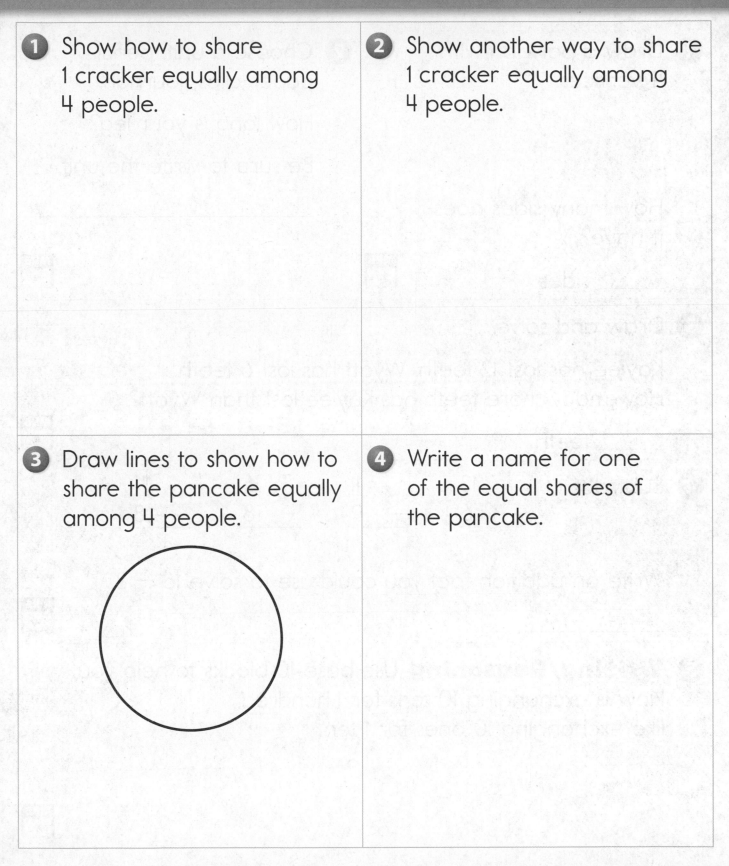

4 Write a name for one of the equal shares of the pancake.

Math Boxes

1 Draw a polygon with 4 vertices.

How many sides does it have?

_____ sides

MRB
127

2 Choose a unit: pencils, paper clips, journals.

How long is your leg?

Be sure to write the unit.

_____

MRB
98

3 Draw and solve.

Kaylee has lost 12 teeth. Wyatt has lost 6 teeth. How many more teeth has Kaylee lost than Wyatt?

_____ teeth

MRB
30

4 Subtract.

$$\begin{array}{cc} 13 \\ -\ 6 \end{array} \qquad \begin{array}{cc} 11 \\ -\ 9 \end{array} \qquad \begin{array}{cc} 16 \\ -\ 8 \end{array} \qquad \begin{array}{cc} 14 \\ -\ 8 \end{array}$$

Write an addition fact you could use to solve $16 - 8$.

_____

MRB
48

5 **Writing/Reasoning** Use base-10 blocks to help you. How is exchanging 10 tens for 1 hundred like exchanging 10 ones for 1 ten?

MRB
70

Sharing Cheese

1) Show how you shared your piece of cheese into two equal shares.

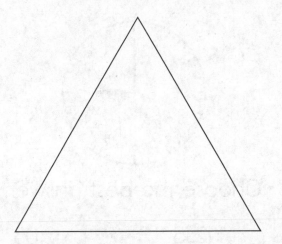

2) Is this piece of cheese divided into two equal shares? _____

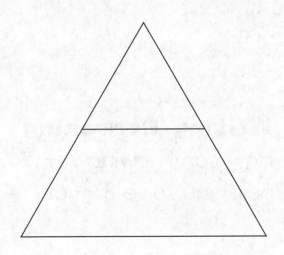

Explain how you know.

_____

_____

_____

Math Boxes

Math Boxes

1 Is this a triangle?

Explain why or why not.

_____

_____

MRB
122-123

2 What time does the clock show?

Choose the best answer.

⬭ 12:15   ⬭ 12:03

⬭ 3:00   ⬭ 3:12

MRB
106

3 Circle the ones digit in each number.

90   9

MRB
73

4 Write the fact family.

Total	
9	
Part	**Part**
9	0

_____

_____

_____

MRB
46

5 **Writing/Reasoning**

How can knowing 7 + 7 help you solve 8 + 6?

MRB
42

166   one hundred sixty-six

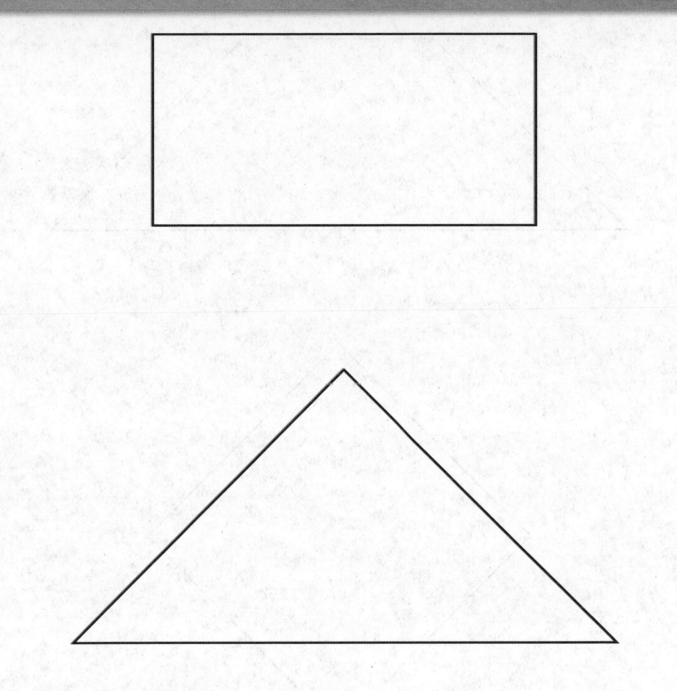

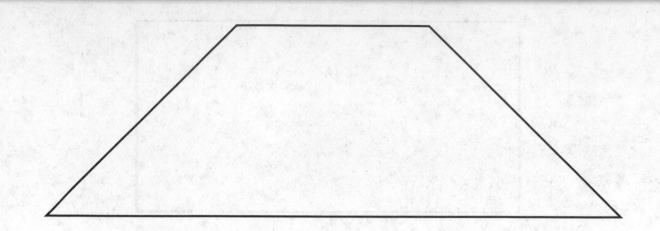

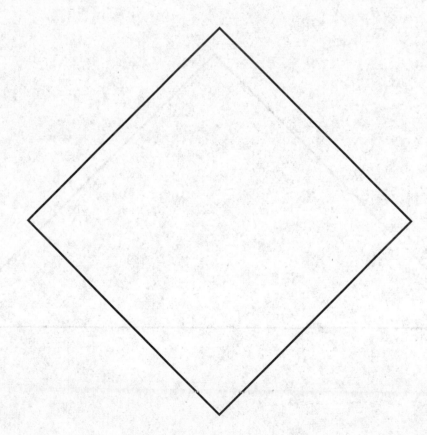

Math Boxes

Math Boxes

1 Name one defining attribute of squares.

MRB
127

2 Divide the circle in half. Then shade 1 half.

MRB
132

3 Subtract.

$16 - \underline{} = 8$

$\underline{} - 0 = 0$

$18 - \underline{} = 9$

$8 - 4 = \underline{}$

Unit

MRB
48

4 Use your number grid. Add.

$\underline{} = 30 + 14$

$\underline{} = 30 + 24$

$\underline{} = 30 + 34$

Unit

MRB
77

5 Subtract.

$\boxed{} = 11 - 6$ $9 - 7 = \boxed{}$ $\boxed{} = 10 - 5$

MRB
49

6 What is the missing number?

in
↓

Rule
− 6

out

in	out
9	3
	7
11	5

Choose the best answer.
- ⬭ 1
- ⬭ 10
- ⬭ 13

MRB
57

Pictures of 3-Dimensional Shapes

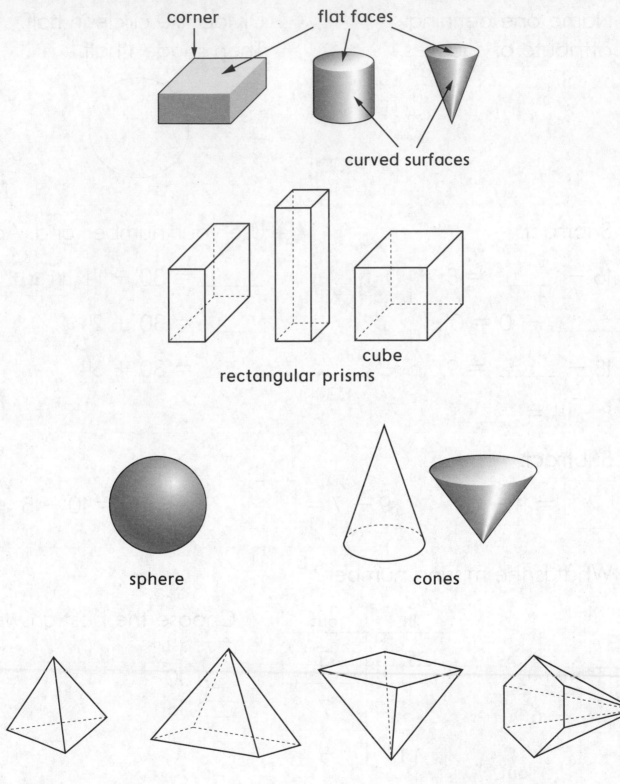

corner

flat faces

curved surfaces

rectangular prisms

cube

sphere

cones

pyramids

Making a Shapes Bar Graph

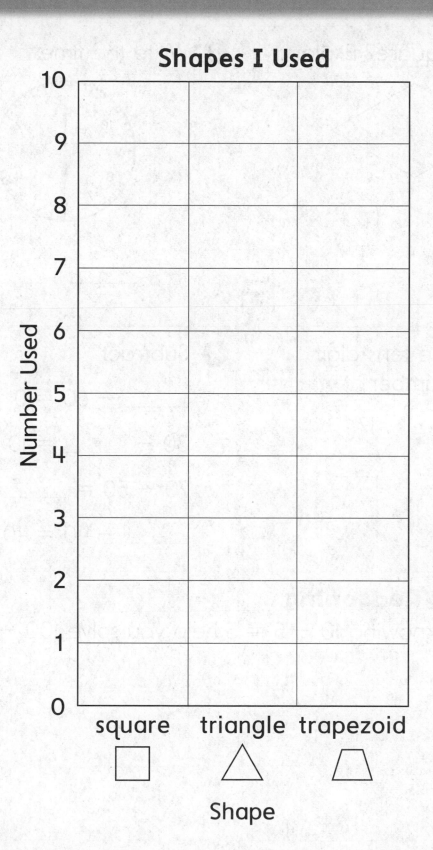

Shapes I Used

Math Boxes

1 Is this a square? Explain.

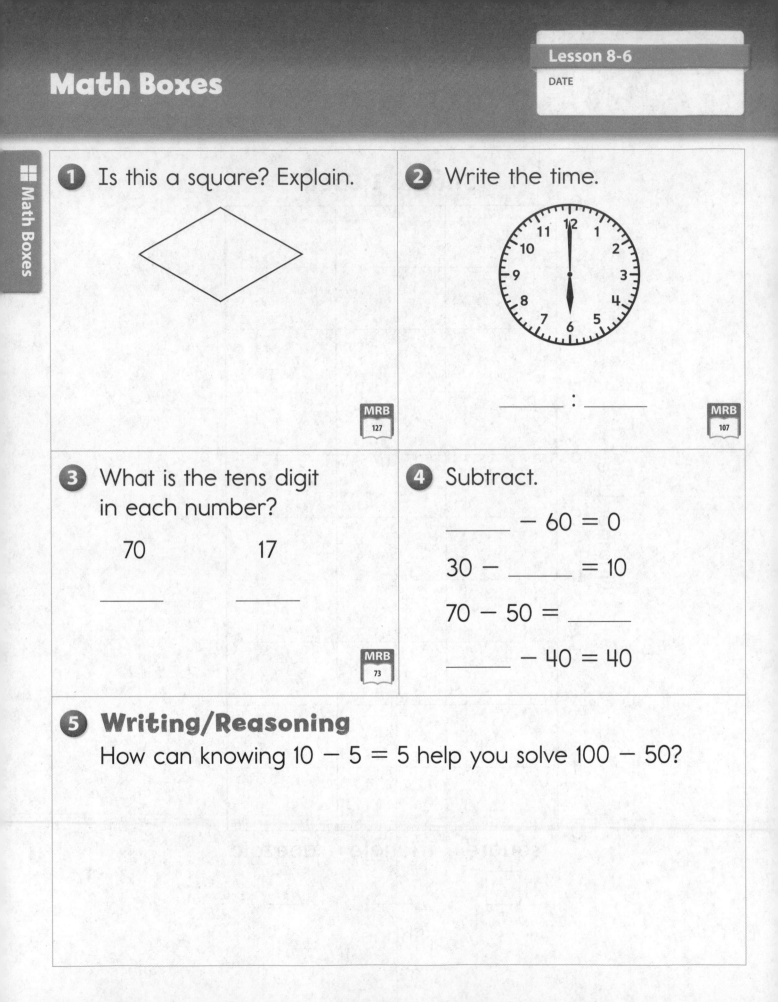

MRB
127

2 Write the time.

_____ : _____

MRB
107

3 What is the tens digit in each number?

70   17

_____   _____

MRB
73

4 Subtract.

_____ − 60 = 0

30 − _____ = 10

70 − 50 = _____

_____ − 40 = 40

5 **Writing/Reasoning**
How can knowing 10 − 5 = 5 help you solve 100 − 50?

Building with 3-Dimensional Shapes Record Sheet

Draw what you made.
Label the 3-dimensional shapes you used.

Sorting by Strategies

Record four facts that you could solve using near doubles.

Record four facts that you could solve by making 10.

Near-doubles strategy	Making-10 strategy

List a fact that you could solve using *either* strategy:

_____

Explain how you could solve it using the near-doubles strategy.

_____

_____

Explain how you could solve it using the making-10 strategy.

_____

_____

Drawing and Naming Equal Shares

① Divide the circle into 2 equal shares.

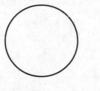

② Write a name for one share. _____

③ Write a name for both shares. _____

④ Divide the circle into 4 equal shares.

⑤ Write a name for one share. _____

⑥ Which is bigger, 1 half of the circle or 1 fourth of the same circle? Why?

_____

_____

Try This

⑦ Which is smaller, 1 half of a rectangle or 1 quarter of the same rectangle? Why?

_____

_____

Math Boxes

Math Boxes

1 Circle all the attributes the 2 shapes have in common.

They have straight sides.
They have square corners.
They have curved sides.
They are shaded.

MRB
128

2 Draw hands to show 7 o'clock.

MRB
106-107

3 Find the rule.

Rule

56   66   76   86

4 Use <, >, or =.

5 Subtract.

Unit

$9 - 8 =$ _____

$90 - 80 =$ _____

_____ $= 5 - 4$

_____ $= 50 - 40$

6 Solve.

Unit

$9 +$ _____ $= 10$

☐ $- 2 = 6$

☐ $= 7 - 6$

$5 = 4 +$ _____

MRB
49

Math Boxes
Preview for Unit 9

Math Boxes

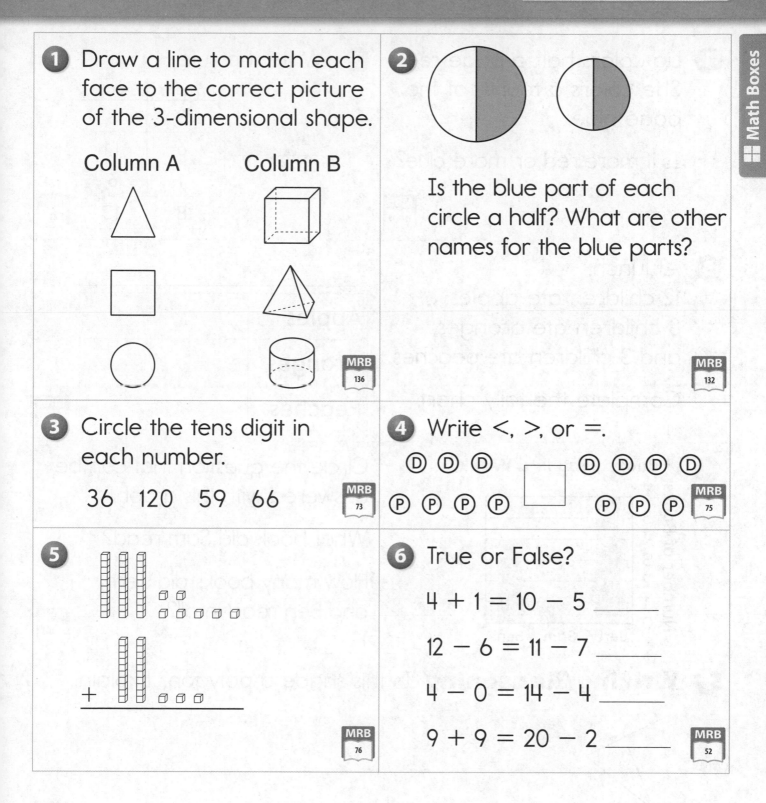

① Draw a line to match each face to the correct picture of the 3-dimensional shape.

Column A   Column B

MRB 136

② Is the blue part of each circle a half? What are other names for the blue parts?

MRB 132

③ Circle the tens digit in each number.

36   120   59   66

MRB 73

④ Write <, >, or =.

Ⓓ Ⓓ Ⓓ ☐ Ⓓ Ⓓ Ⓓ Ⓓ

Ⓟ Ⓟ Ⓟ Ⓟ   Ⓟ Ⓟ Ⓟ

MRB 75

⑤

MRB 76

⑥ True or False?

4 + 1 = 10 − 5 _____

12 − 6 = 11 − 7 _____

4 − 0 = 14 − 4 _____

9 + 9 = 20 − 2 _____

MRB 52

Math Boxes

① Lia colors half a page red. She colors a fourth of the page blue.

Is it more red or more blue?

_____

MRB
132-133

② Write the rule.

in → Rule → out

in	out
3	8
6	11
1	6
8	13

MRB
58

③ At lunch,
12 children ate apples,
8 children ate oranges,
and 3 children ate peaches.

Complete the tally chart.

Apples	
Oranges	
Peaches	

MRB
113

④ **Books Read in a Week**

Number of Books: 6, 5, 4, 3, 2, 1, 0

Jerry   Sam   Ben

Circle the question that can be answered with this graph.

What book did Sam read?

How many books did Jerry and Ben read in all?

MRB
116

⑤ **Writing/Reasoning** Is this shape a polygon? Explain.

MRB
125

Number-Grid Puzzles 1

Solve the puzzles.

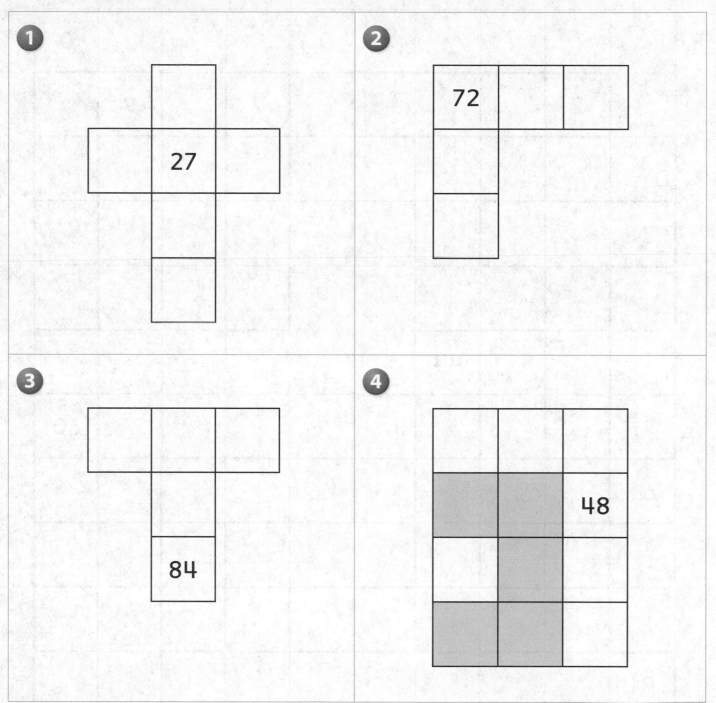

1 27

2 72

3 84

4 48

Number-Grid Puzzles 2

Fill in the missing numbers.

									0
	2					7		9	
21									
			44						
									60
61					66				
91									

Math Boxes

1 Name 2 things that are the same about these shapes.

MRB
127

2 Show half past 11 o'clock on the clocks.

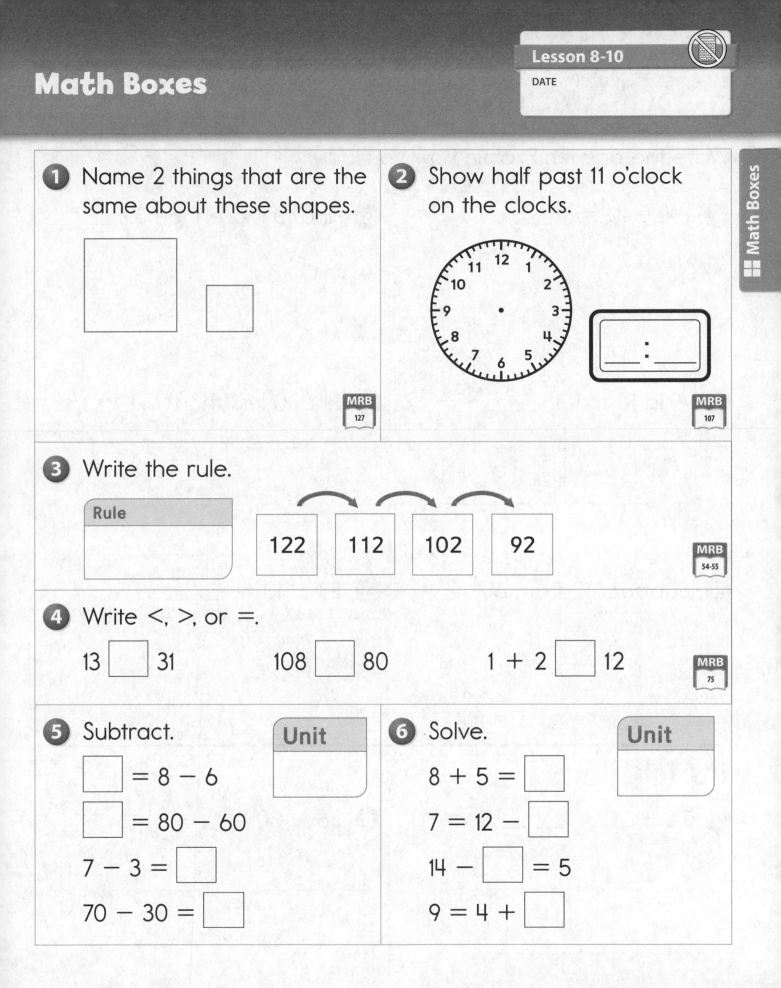

MRB
107

3 Write the rule.

Rule

122   112   102   92

MRB
54-55

4 Write <, >, or =.

13 ☐ 31   108 ☐ 80   1 + 2 ☐ 12

MRB
75

5 Subtract.

Unit

☐ = 8 − 6

☐ = 80 − 60

7 − 3 = ☐

70 − 30 = ☐

6 Solve.

Unit

8 + 5 = ☐

7 = 12 − ☐

14 − ☐ = 5

9 = 4 + ☐

one hundred eighty-one   181

Solving 10 More, 10 Less Problems

Write the answer. Explain how you know.

1 84 − 10 = _____

2 Find 10 more than 39. _____

3 Add 10 to 42. _____

4 Find 10 less than 97. _____

5 Subtract 10 from 61. _____

6 23 + 10 = _____

Try This

7 75 + 20 = _____

8 56 − 30 = _____

Applying and Finding Rules

Fill in the missing numbers.

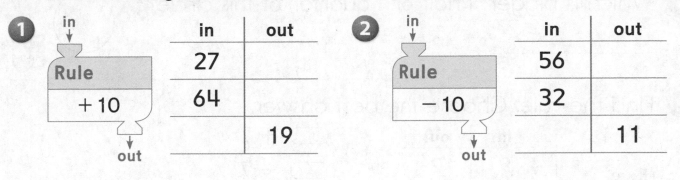

1

Rule +10

in	out
27	
64	
	19

2

Rule −10

in	out
56	
32	
	11

Find the rule.
Fill in the missing numbers.

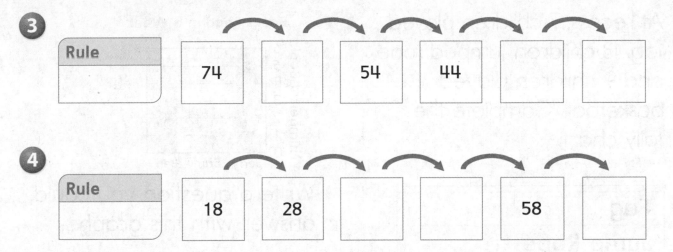

3

Rule

74		54	44		

4

Rule

18	28			58	

5 Explain how you found the rule for Problem 4 in your head.

_____

_____

_____

_____

Math Boxes

1 Which is bigger: 1 half or 1 quarter of this circle?

_____

MRB 132-133

2 Find the rule. Choose the best answer.

in

Rule

out

in	out
9	2
14	7
15	8

◯ + 7
◯ − 7
◯ + 5
◯ + 1

MRB 58

3 At recess, 7 children played tag, 13 children jumped rope, and 9 children played basketball. Complete the tally chart.

Tag	
Jump Rope	
Basketball	

MRB 113

4 **Books Read in a Week**

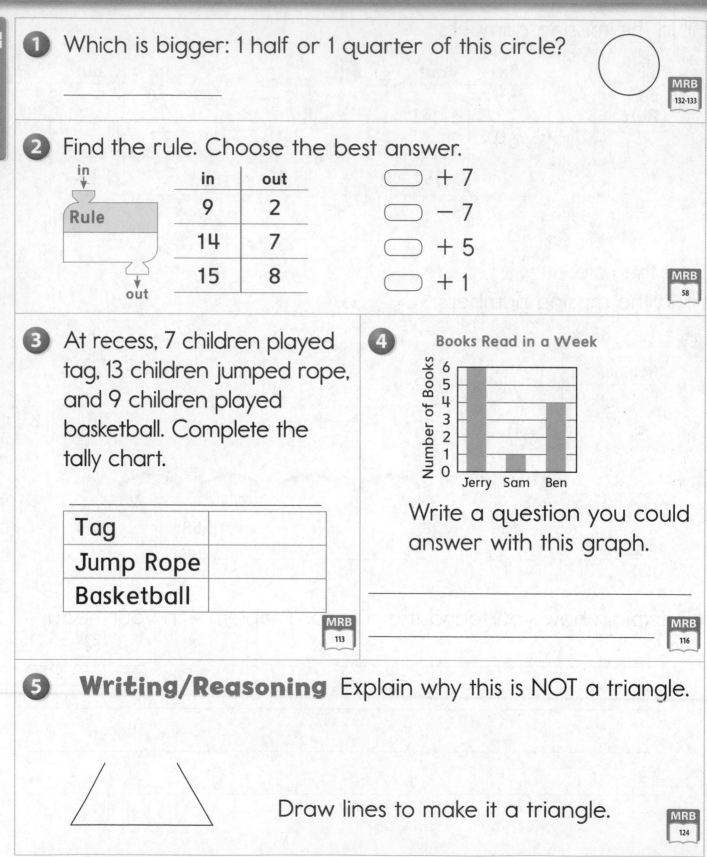

Number of Books

6 5 4 3 2 1 0

Jerry Sam Ben

Write a question you could answer with this graph.

_____

_____

MRB 116

5 **Writing/Reasoning** Explain why this is NOT a triangle.

Draw lines to make it a triangle.

MRB 124

184   one hundred eighty-four

Math Boxes
Preview for Unit 9

Math Boxes

1 Draw a line to match each face to the correct picture of the 3-dimensional shape.

Column A   Column B

MRB 136

2 Each part is half.

What could you call both shares? _____

MRB 132

3 Circle the ones digit.

22   58   100

4   60   16

MRB 73

4 Write <, >, or =.

Ⓓ Ⓓ   Ⓓ Ⓓ Ⓓ
Ⓓ Ⓓ Ⓓ   Ⓟ Ⓟ Ⓟ
Ⓟ Ⓟ Ⓟ _____ Ⓟ Ⓟ Ⓟ
Ⓟ Ⓟ   Ⓟ Ⓟ

MRB 75

5

+

MRB 76

6 Is the number sentence true? _____

If not, change only one number or symbol to make it true.

$3 + 3 + 5 = 12$

MRB 52

one hundred eighty-five   185

Measuring with Rulers

1 How wide is your *Math Journal*?
Measure by lining up **paper clips.**

My journal is about _____ paper-clips wide.

Measure again using your **paper-clip ruler.**
Did you get the same answer? Explain why or why not.

_____

_____

Measure more objects.
Use only your paper-clip ruler.

2 I measured _____.

It is about _____ paper-clip units long.

3 I measured _____.

It is about _____ paper-clip units long.

4 I measured _____.

It is about _____ paper-clip units long.

5 I measured _____.

It is about _____ paper-clip units long.

Math Boxes

Math Boxes

① Draw a polygon with 6 sides.

MRB
124

② Color the rectangles.

MRB
128

③ How much of the shape is colored?

_____

MRB
132

④ How many equal shares are there in each shape?

_____   _____

Color one of the larger shares.

MRB
132-133

⑤ You have 40¢. You buy a pencil for 30¢.

Unit

How much do you have left? _____

Number model:

_____

MRB
28

⑥ Find the rule and the missing numbers.

in	out
12	10
	0
6	

in
↓

Rule

↓

out

MRB
56-58

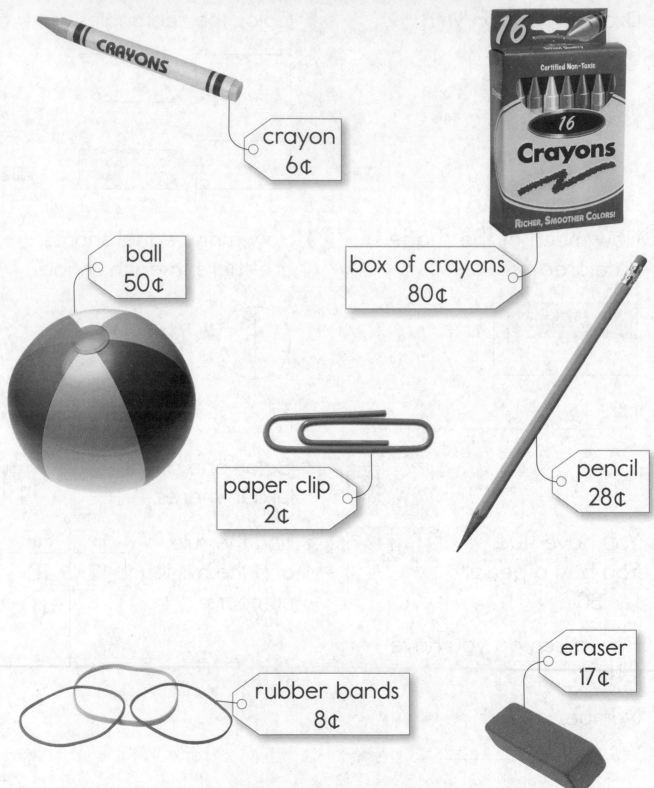

crayon
6¢

box of crayons
80¢

ball
50¢

paper clip
2¢

pencil
28¢

rubber bands
8¢

eraser
17¢

(clockwise from top) ©McGraw-Hill Educat on/Ken Cavanagh; ©McGraw-Hill Education; ©Thinkstock Images/Getty Images;
Burke/Triolo/Brand X Pictures/Jupiterimages; ©Ryan McVay/Getty Images; ©McGraw-Hill Education; ©Jacques Cornell;
©McGraw-Hill Education/Dot Box Inc.

School Store Mini-Poster 2

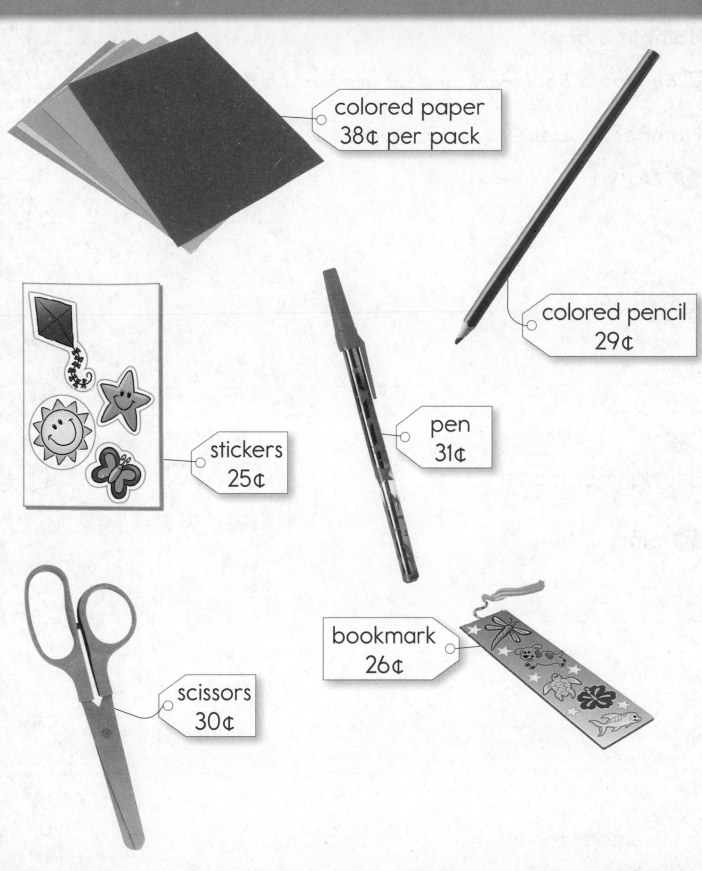

colored paper
38¢ per pack

colored pencil
29¢

stickers
25¢

pen
31¢

scissors
30¢

bookmark
26¢

Recording Number Stories

Sample Story

I bought a bookmark and an eraser. I paid 43¢ in all.

Number Model: 26¢ + 17¢ = 43¢

 1 Story 1

Number model: _____

 2 Story 2

Number model: _____

Math Boxes

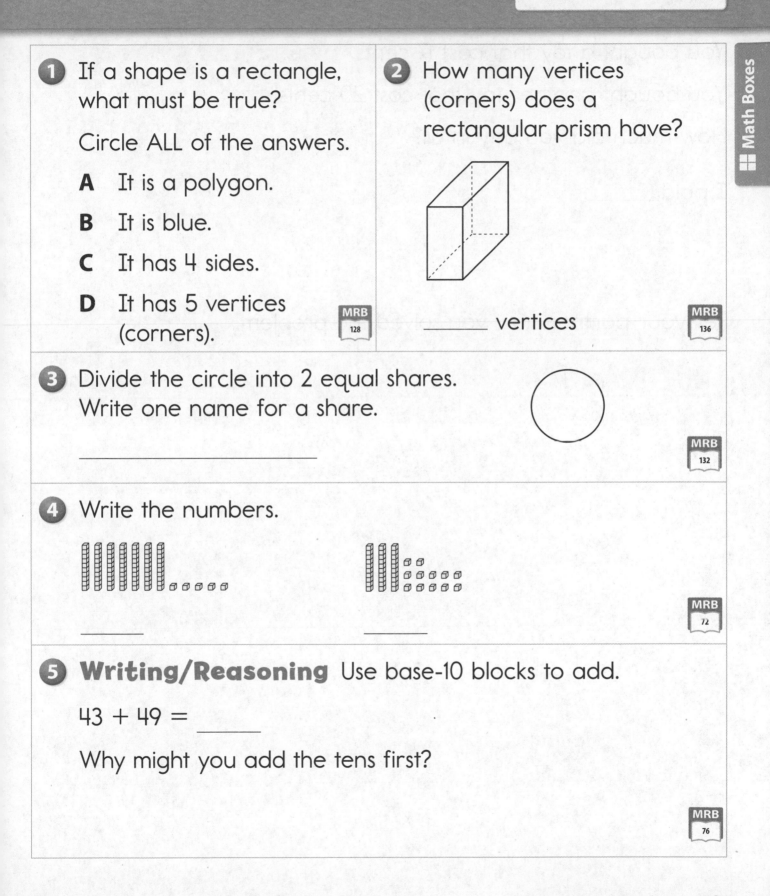

1. If a shape is a rectangle, what must be true?

   Circle ALL of the answers.

   A It is a polygon.

   B It is blue.

   C It has 4 sides.

   D It has 5 vertices (corners).

   MRB 128

2. How many vertices (corners) does a rectangular prism have?

   _____ vertices

   MRB 136

3. Divide the circle into 2 equal shares. Write one name for a share.

   _____

   MRB 132

4. Write the numbers.

   _____   _____

   MRB 72

5. **Writing/Reasoning** Use base-10 blocks to add.

   43 + 49 = _____

   Why might you add the tens first?

   MRB 76

Buying Toys

You bought a toy that cost 17 cents.

You bought another toy that cost 20 cents.

How much did you pay in all?

I paid _____.

Tell your partner how you solved the problem.

Math Boxes

1 Use 2 triangles to make a new shape. Draw it here. Use your Pattern-Block Template.

Is it a polygon?_____

MRB
129

2 Draw a polygon with 4 sides that are the same length.

MRB
127-128

3 Circle all the names for the colored region.

1 fourth   1 of 2 equal pieces

1 half   whole

MRB
132-133

4 How many equal pieces are there in each shape?

_____   _____

Color one of the smaller pieces.

MRB
132-133

5 You have 70¢.
You buy a pen for 40¢.

How much do you have left?_____¢

Number model:

_____

MRB
28

6 Find the rule.

in

Rule

out

Write a true equation to check the rule.

9 _____ _____ = 17

in	out
3	11
9	17
5	13
10	18

MRB
56-58

Broken-Calculator Puzzles

Use + and − to solve the broken-calculator puzzles.
Use a calculator to check your answers.

 Imagine your 1-key is broken.
Write at least 3 ways to show 11 without using the 1-key.

 Imagine your 2-key is broken.
Write at least 3 ways to show 20 without using the 2-key.

 Imagine your 4-key is broken.
Write at least 3 ways to show 34 without using the 4-key.

 Write your own puzzle. Give it to your partner to solve.

Telling Time

Record the times on the digital clocks.

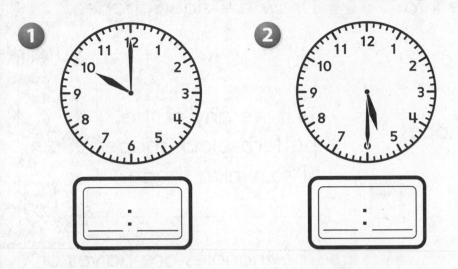

Draw clock hands to show the times.

7 Show the time on both clocks.

half-past seven o'clock

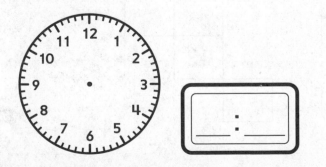

Math Boxes

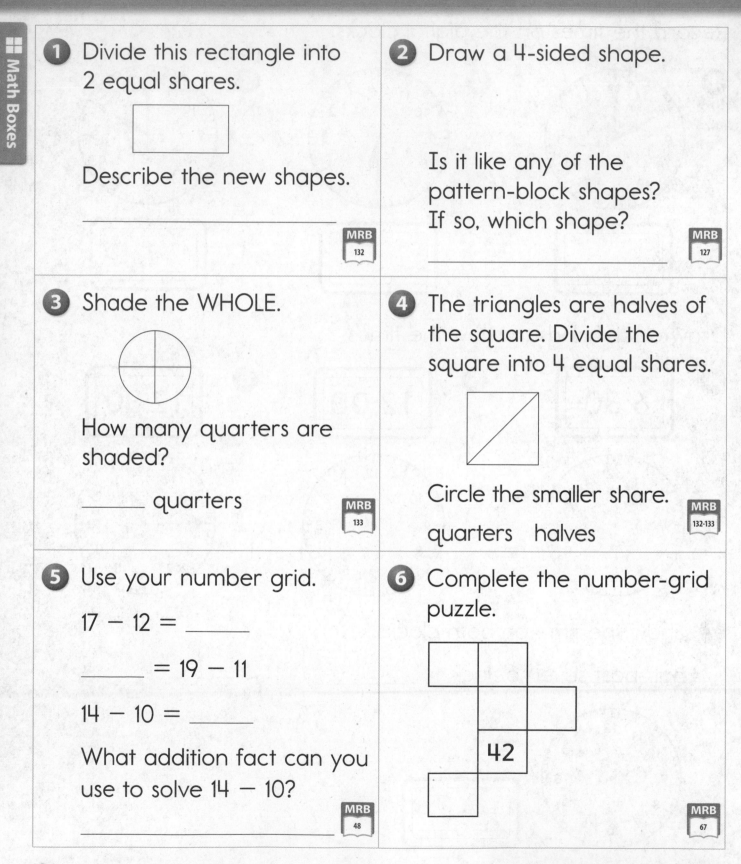

1 Divide this rectangle into 2 equal shares.

Describe the new shapes.

_____

MRB 132

2 Draw a 4-sided shape.

Is it like any of the pattern-block shapes? If so, which shape?

_____

MRB 127

3 Shade the WHOLE.

How many quarters are shaded?

_____ quarters

MRB 133

4 The triangles are halves of the square. Divide the square into 4 equal shares.

Circle the smaller share.

quarters   halves

MRB 132-133

5 Use your number grid.

$17 - 12 =$ _____

_____ $= 19 - 11$

$14 - 10 =$ _____

What addition fact can you use to solve $14 - 10$?

_____

MRB 48

6 Complete the number-grid puzzle.

42

MRB 67

196   one hundred ninety-six

Turning Rectangles into Other Shapes

Cut the rectangles into triangles.
Glue the triangles to show your answers.
The shapes you make must fit in the boxes.

1 Use two triangles to make one large triangle.

2 Use two triangles to make a shape with 4 sides.
The shape should NOT be a rectangle.

3 Use two triangles to make a shape with 5 sides.

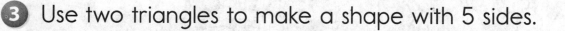

Math Boxes

1 Name at least 2 attributes of all trapezoids.

MRB
127

2 What is a defining attribute of a rectangular prism?

MRB
134

3 Divide the rectangle into 4 equal shares.

Write a name for one share.

_____

MRB
133

4 Write the numbers.

_____

_____

MRB
72

5 **Writing/Reasoning** Use your number grid or base-10 blocks to solve. Explain what you do first.

$35 + \boxed{} = 76$

MRB
76-77

one hundred ninety-nine   199

Vending Machine Prices

These snacks went up in price. How much did each price go up?

_____Crackers_____ _20_ ¢   _____ _____ ¢

_____ _____ ¢   _____ _____ ¢

_____ _____ ¢   _____ _____ ¢

Niko has 30¢. Leela has 35¢. Wei has 50¢.
Niko, Leela, and Wei each buy a package
of cheese sticks.

How much money does each child have left?

Niko will have _____ ¢ left. Number model: _____

Leela will have _____ ¢ left. Number model: _____

Wei will have _____ ¢ left. Number model: _____

Try This

Niko, Leela, and Wei put together all of the money
they have left. How much do they have in all? _____ ¢

Number model: _____

Creating Vending Machine Number Stories

Write number stories about the vending machine
on journal page 197.

Trade with a partner.
Solve the number stories and then write a number model.
Trade again and check your partner's answers.

1

2

3

Name-Collection Boxes

 1 Write four different names for 19.

19	9 + 9 + 1

Write a true number sentence for each name you added.

$9 + 9 + 1 = 19$

_____

_____

_____

2 Cross out the names that do not belong.

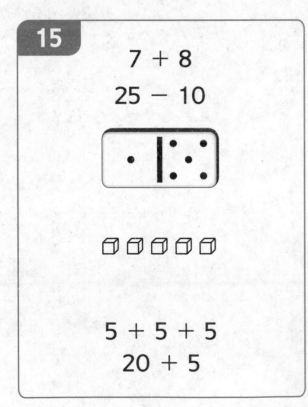

15	

7 + 8

25 − 10

5 + 5 + 5

20 + 5

Write a true number sentence for each name that does belong.

_____

_____

_____

_____

Math Boxes

Math Boxes

1 Divide the rectangles into 2 equal parts in different ways.

Describe the shapes you made.

MRB
132

2 Name one attribute of this polygon.

_____

MRB
127

3 Shade the WHOLE.

What else could you call the shaded part?

_____

MRB
133

4 Alex shares a tart equally with 1 friend.

Walter shares a tart of the same size with 3 friends.

Whose piece is bigger?

_____

MRB
132-133

5 Solve on your number grid.

_____ = 95 − 86

What addition number sentence is the same as _____ = 95 − 86?

MRB
82

6 Complete the number-grid puzzle.

79

MRB
67

Silly Animal Stories

Example:

Unit
inches

koala
24 in.

penguin
36 in.

How tall are the koala and penguin all together?

24 + 36 = 60

60 inches

 Silly Story

Unit

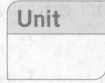

---

2 **Silly Story**

Unit

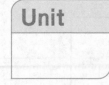

Math Boxes

1 How are squares different from other rectangles?

Choose the best answer.

- ⬭ They are polygons.
- ⬭ They are blue.
- ⬭ They have 3 sides.
- ⬭ The sides are all the same length.

MRB 128

2 Shade 1 quarter of the rectangle.

MRB 133

3 Divide one circle in half. Divide one circle in fourths.

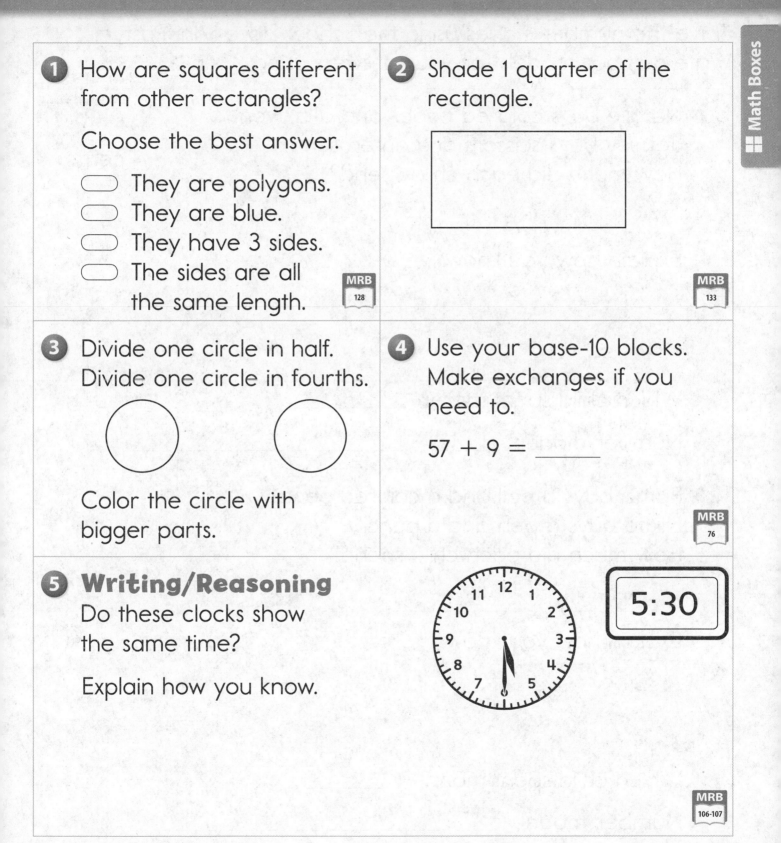

Color the circle with bigger parts.

4 Use your base-10 blocks. Make exchanges if you need to.

57 + 9 = _____

MRB 76

5 **Writing/Reasoning**
Do these clocks show the same time?

Explain how you know.

5:30

MRB 106-107

Adding and Comparing Sums

Solve the number stories using the School Store Mini-Posters.
Write number models for your answers using < or >.

---

1 Natalie buys colored paper and a crayon.
Jeffrey buys scissors and an eraser.
How much did each child spend?

Unit

cents

Explain how you know.

Which child spent more? _____

Number model: _____

---

2 Ramzi buys a ball and a colored pencil.
Nikita buys a pen and a pencil.
How much did each child spend?

Explain how you know.

Which child spent more? _____

Number model: _____

Making Smallest and Largest Numbers

1. Shuffle number cards 1–9. Put them facedown on your desk.

2. Draw two number cards.

3. Make the smallest number you can.

4. Make the largest number you can.

5. Record the numbers.

6. Put the cards back and do it again.

Turn	Cards	Smaller Number	Larger Number
Example	5, 3	35	53
1			
2			
3			

Use the numbers in the table to write a number sentence for each row using <, >, and =.

1. _____

2. _____

3. _____

Try This

What is the smallest number you can make with your cards? _____ The largest? _____

Math Boxes

① Draw two different circles.

What is different?

MRB
122-123

② Find a base-10 cube.

How many faces does a cube have? _____ faces

How many vertices does a cube have? _____ vertices

MRB
136

③ Color 1 half of each rectangle a different way.

MRB
132

④ Circle the longer object:
base-10 cube   paper clip

Measure the length of your journal with both.

The length of my journal is _____ base-10 cubes or

_____ paper clips.

MRB
96,98

⑤ Circle the numbers that are less than:

13   113

23   14

MRB
74

⑥ Fill in the missing numbers.

in
↓

Rule

+ 10

↓
out

in	out
95	105
60	
	112
76	

MRB
56-58

Using Different Strategies

Solve the number sentences below. Record the tool or strategy you used and explain why you chose it.

• number grid	• base-10 blocks
• counting back	• fact extensions
• making 10	• counting up
• changing to an easier number	• 10 more or 10 less in my head
• number line	• combining place-value groups

Example: $20 + 36 =$ ___56___

I used a number grid because moving down 2 rows is an easy way to add 20. I started at 36 and moved down two rows.

1 $58 + 3 =$ _____

_____

_____

2 $17 + 30 =$ _____

_____

_____

3 $79 + 14 =$ _____

_____

_____

4 $6 + 46 + 4 =$ _____

_____

_____

Math Boxes

① Circle the rectangles.
Shade the squares.

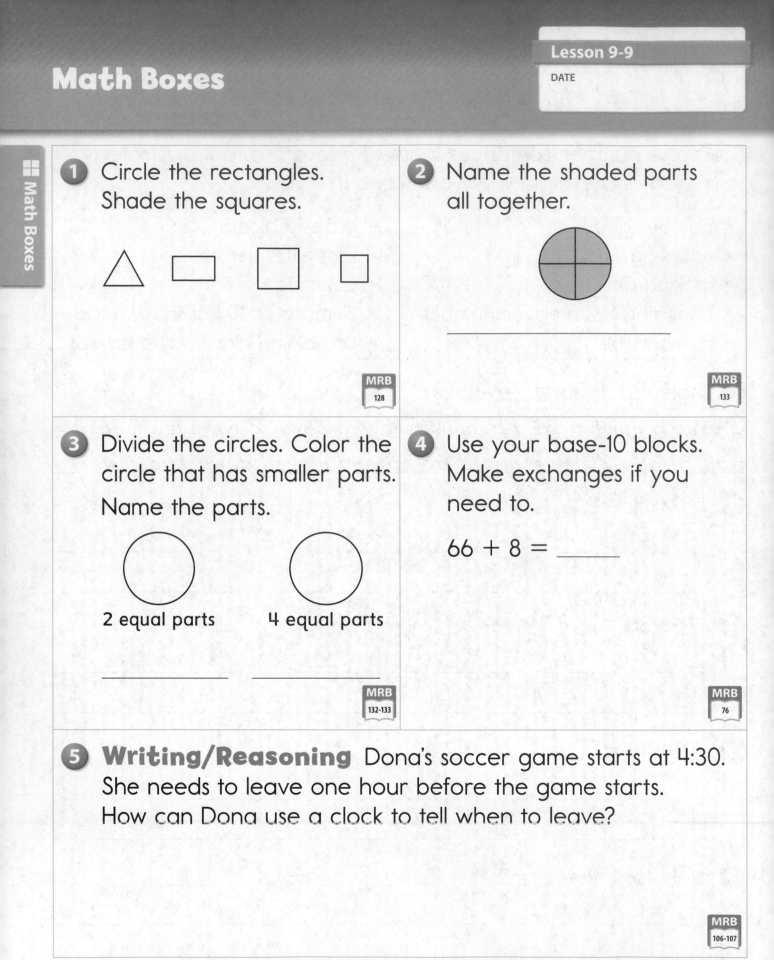

MRB
128

② Name the shaded parts
all together.

_____

MRB
133

③ Divide the circles. Color the
circle that has smaller parts.
Name the parts.

2 equal parts   4 equal parts

_____   _____

MRB
132-133

④ Use your base-10 blocks.
Make exchanges if you
need to.

66 + 8 = _____

MRB
76

⑤ **Writing/Reasoning** Dona's soccer game starts at 4:30.
She needs to leave one hour before the game starts.
How can Dona use a clock to tell when to leave?

MRB
106-107

Reviewing Attributes of 3-Dimensional Shapes

Write a defining attribute for each shape.

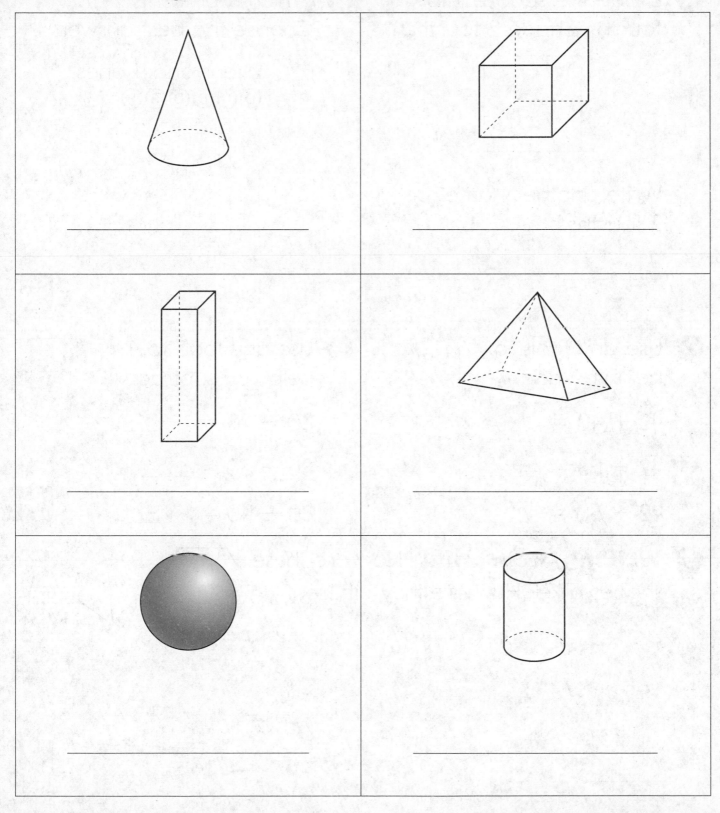

Math Boxes

1 Divide the square into fourths. Shade 2 fourths.

Write another name for 2 fourths.

_____

MRB
133

2 Which represents 24? Choose the best answer.

⬭ 2 tens and 4 ones

⬭ ⒟⒟⒟⒟ⓅⓅ

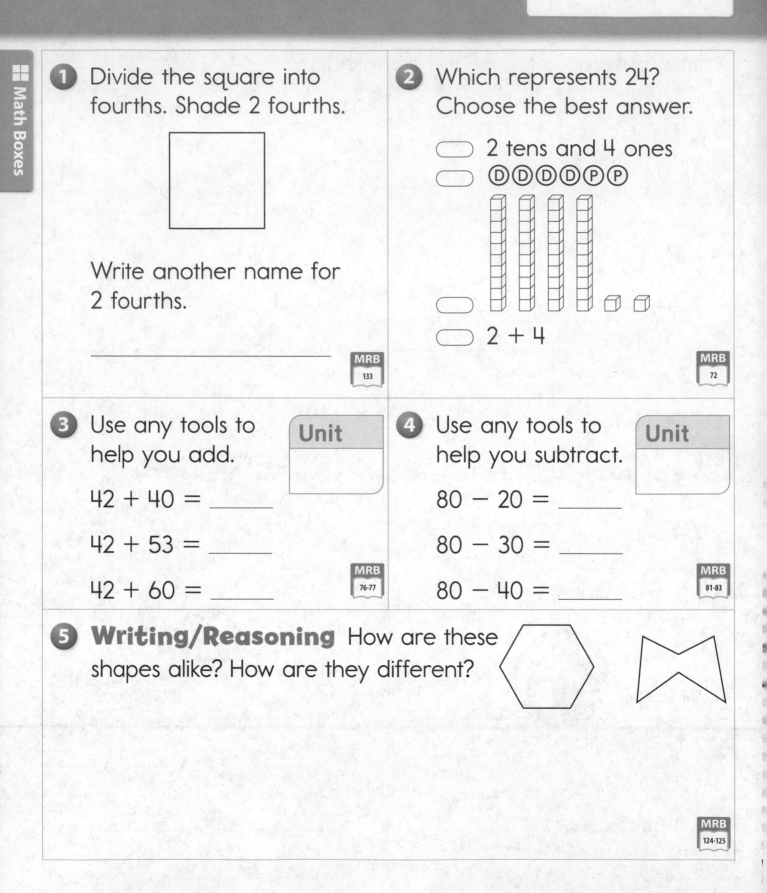

⬭

⬭ 2 + 4

MRB
72

3 Use any tools to help you add.

Unit

42 + 40 = _____

42 + 53 = _____

42 + 60 = _____

MRB
76-77

4 Use any tools to help you subtract.

Unit

80 − 20 = _____

80 − 30 = _____

80 − 40 = _____

MRB
81-83

5 **Writing/Reasoning** How are these shapes alike? How are they different?

MRB
124-125

Partitioning Granola Squares

1. Show two different ways to share a granola square among 4 people.

2. Write two different names for 1 of the shares.

_____

_____

3. Sasha ate all 4 shares of the granola square. Name the share she ate.

_____

4. Circle which is bigger.

1 half of a granola square   1 fourth of a granola square

Why?_____

_____

Math Boxes

1 Draw a polygon.

Describe your polygon.

MRB
124-125

2 How many vertices does a cylinder have?

_____ vertices

How many vertices does a rectangular prism have?

_____ vertices

MRB
136

3 Shade 1 quarter of each shape.

MRB
133

4 Circle the longer object.

your pencil   your crayon

Measure the height of your desk with both.

The height of my desk is _____ pencils or _____ crayons.

MRB
96, 98

5 Write three numbers that are more than 80.

MRB
74

6 Fill in the missing numbers.

in
↓

Rule

+ 10

out

in	out
14	24
98	
	52
53	

MRB
56-58

Math Boxes

1 Write 2 names for the shaded part.

_____

_____

MRB
132

2 Show 50¢ two ways.
Use | and ▪ , ⒟ and ⓟ, or drawings.

MRB
72

3 Add. Use tools if you like.

Unit

$45 +$ _____ $= 75$

_____ $+ 34 = 75$

_____ $+ 28 = 75$

MRB
76-77

4 Subtract. Use tools if you like.

Unit

$$\begin{array}{ccc} 90 & 80 & 70 \\ -20 & -30 & -50 \\ \hline \end{array}$$

MRB
81-83

5 **Writing/Reasoning** How are triangles and squares alike? How are they different?

MRB
124-125

two hundred fifteen **215**

Notes

My Facts Inventory Record, Part 1

Addition Fact	Know It	Don't Know It	How I Can Figure It Out...
0 + 1			
7 + 2			
0 + 3			
3 + 2			
0 + 5			
10 + 2			
3 + 1			
2 + 2			
6 + 0			
1 + 1			

My Facts Inventory Record, Part 1 (continued)

DATE

Addition Fact	Know It	Don't Know It	How I Can Figure It Out...
4 + 1			
5 + 2			
4 + 0			
0 + 2			
1 + 5			
7 + 0			
10 + 1			
2 + 4			
1 + 6			

Addition Fact	Know It	Don't Know It	How I Can Figure It Out...
0 + 9			
1 + 2			
8 + 1			
0 + 0			
9 + 2			
2 + 6			
8 + 0			
7 + 1			
0 + 10			

My Facts Inventory Record, Part 2

DATE

Addition Fact	Know It	Don't Know It	How I Can Figure It Out...
10 + 10			
5 + 5			
10 + 5			
8 + 8			
7 + 10			
3 + 3			
8 + 2			
4 + 10			
6 + 4			

My Facts Inventory Record, Part 2 (continued)

DATE

Addition Fact	Know It	Don't Know It	How I Can Figure It Out...
10 + 8			
7 + 7			
6 + 6			
3 + 10			
9 + 9			
1 + 9			
7 + 3			
6 + 10			
4 + 4			
10 + 9			

My Facts Inventory Record, Part 3

DATE

Addition Fact	Know It	Don't Know It	How I Can Figure It Out…
3 + 4	✓		
7 + 8	✓		
4 + 6	✓		
7 + 9	✓	✗	///////// ////
6 + 5			
9 + 8			
4 + 5			
7 + 6			
3 + 5			
5 + 7			

My Facts Inventory Record, Part 4

DATE

Addition Fact	Know It	Don't Know It	How I Can Figure It Out...
9 + 7	✓		
5 + 6	✓		
8 + 3	✓		
6 + 9	✓		
7 + 5			
4 + 9		✗	
8 + 6			
7 + 4			
9 + 5			
4 + 8			
6 + 3			
8 + 7			
9 + 3			
5 + 8			

Fact Triangles 4

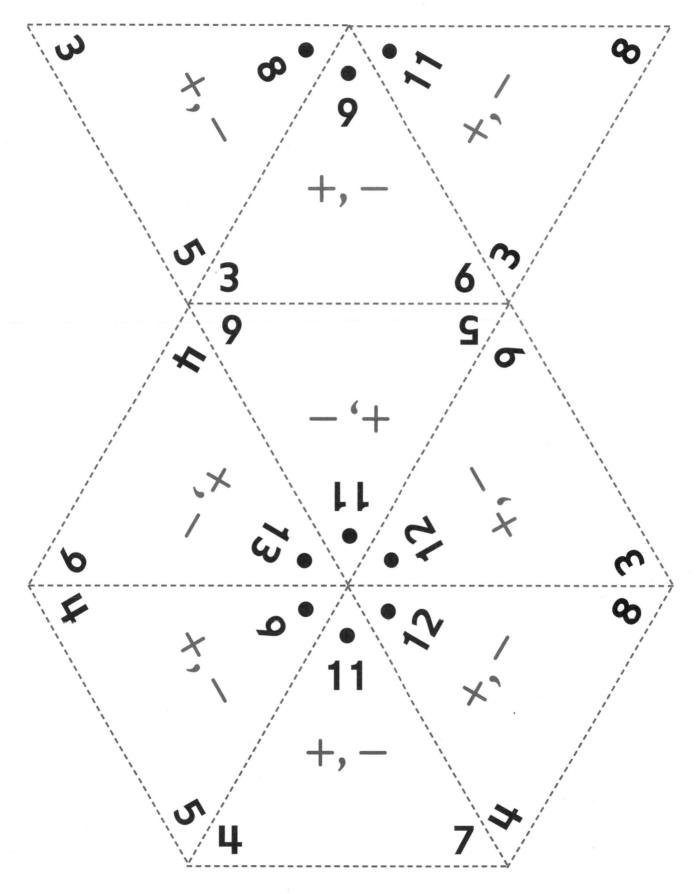

Fact Triangles 5

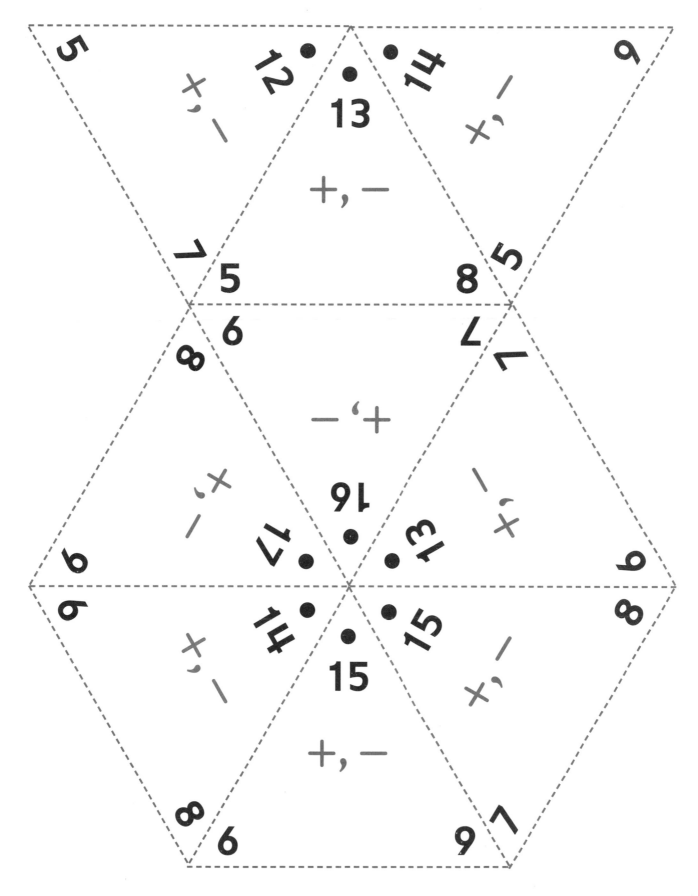

Fact Triangles 6

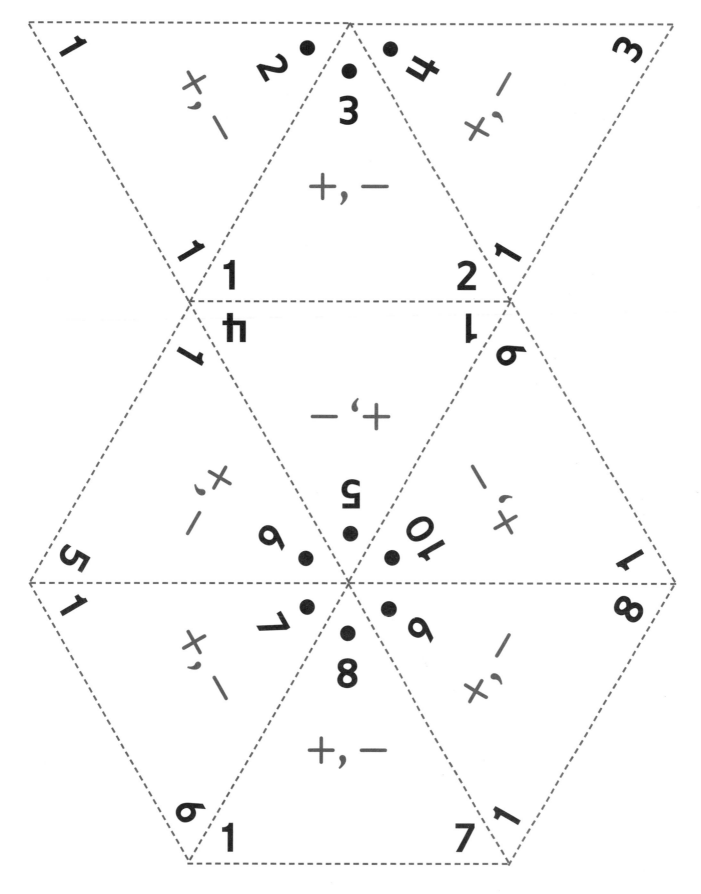